International Political Economy Series

Series Editor: **Timothy M. Shaw**, Visiting Professor, University of Massachusetts Boston, USA and Emeritus Professor, University of London, UK

The global political economy is in flux as a series of cumulative crises impacts its organization and governance. The IPE series has tracked its development in both analysis and structure over the last three decades. It has always had a concentration on the global South. Now the South increasingly challenges the North as the centre of development, also reflected in a growing number of submissions and publications on indebted Eurozone economies in Southern Europe.

An indispensable resource for scholars and researchers, the series examines a variety of capitalisms and connections by focusing on emerging economies, companies and sectors, debates and policies. It informs diverse policy communities as the established trans-Atlantic North declines and 'the rest', especially the BRICS, rise.

Titles include:

Monique Taylor
THE CHINESE STATE, OIL AND ENERGY SECURITY

Benedicte Bull, Fulvio Castellacci and Yuri Kasahara
BUSINESS GROUPS AND TRANSNATIONAL CAPITALISM IN CENTRAL AMERICA
Economic and Political Strategies

Leila Simona Talani
THE ARAB SPRING IN THE GLOBAL POLITICAL ECONOMY

Andreas Nölke (*editor*)
MULTINATIONAL CORPORATIONS FROM EMERGING MARKETS
State Capitalism 3.0

Roshen Hendrickson
PROMOTING U.S. INVESTMENT IN SUB-SAHARAN AFRICA

Bhumitra Chakma
SOUTH ASIA IN TRANSITION
Democracy, Political Economy and Security

Greig Charnock, Thomas Purcell and Ramon Ribera-Fumaz
THE LIMITS TO CAPITAL IN SPAIN
Crisis and Revolt in the European South

Felipe Amin Filomeno
MONSANTO AND INTELLECTUAL PROPERTY IN SOUTH AMERICA

Eirikur Bergmann
ICELAND AND THE INTERNATIONAL FINANCIAL CRISIS
Boom, Bust and Recovery

Yildiz Atasoy (*editor*)
GLOBAL ECONOMIC CRISIS AND THE POLITICS OF DIVERSITY

Gabriel Siles-Brügge
CONSTRUCTING EUROPEAN UNION TRADE POLICY
A Global Idea of Europe

Jewellord Singh and France Bourgouin (*editors*)
RESOURCE GOVERNANCE AND DEVELOPMENTAL STATES IN THE GLOBAL SOUTH
Critical International Political Economy Perspectives

Tan Tai Yong and Md Mizanur Rahman (*editors*)
DIASPORA ENGAGEMENT AND DEVELOPMENT IN SOUTH ASIA

Leila Simona Talani, Alexander Clarkson and Ramon Pacheco Pardo (*editors*)
DIRTY CITIES
Towards a Political Economy of the Underground in Global Cities

Matthew Louis Bishop
THE POLITICAL ECONOMY OF CARIBBEAN DEVELOPMENT

Xiaoming Huang (*editor*)
MODERN ECONOMIC DEVELOPMENT IN JAPAN AND CHINA
Developmentalism, Capitalism and the World Economic System

Bonnie K. Campbell (*editor*)
MODES OF GOVERNANCE AND REVENUE FLOWS IN AFRICAN MINING

Gopinath Pillai (*editor*)
THE POLITICAL ECONOMY OF SOUTH ASIAN DIASPORA
Patterns of Socio-Economic Influence

Rachel K. Brickner (*editor*)
MIGRATION, GLOBALIZATION AND THE STATE

Juanita Elias and Samanthi Gunawardana (*editors*)
THE GLOBAL POLITICAL ECONOMY OF THE HOUSEHOLD IN ASIA

Tony Heron
PATHWAYS FROM PREFERENTIAL TRADE
The Politics of Trade Adjustment in Africa, the Caribbean and Pacific

David J. Hornsby
RISK REGULATION, SCIENCE AND INTERESTS IN TRANSATLANTIC TRADE CONFLICTS

Yang Jiang
CHINA'S POLICYMAKING FOR REGIONAL ECONOMIC COOPERATION

Martin Geiger and Antoine Pécoud (*editors*)
DISCIPLINING THE TRANSNATIONAL MOBILITY OF PEOPLE

Michael Breen
THE POLITICS OF IMF LENDING

Laura Carsten Mahrenbach
THE TRADE POLICY OF EMERGING POWERS
Strategic Choices of Brazil and India

Vassilis K. Fouskas and Constantine Dimoulas
GREECE, FINANCIALIZATION AND THE EU
The Political Economy of Debt and Destruction

Hany Besada and Shannon Kindornay (*editors*)
MULTILATERAL DEVELOPMENT COOPERATION IN A CHANGING GLOBAL ORDER

Caroline Kuzemko
THE ENERGY– SECURITY CLIMATE NEXUS
Institutional Change in Britain and Beyond

Hans Löfgren and Owain David Williams (*editors*)
THE NEW POLITICAL ECONOMY OF PHARMACEUTICALS
Production, Innovation and TRIPS in the Global South

Timothy Cadman (*editor*)
CLIMATE CHANGE AND GLOBAL POLICY REGIMES
Towards Institutional Legitimacy

International Political Economy Series
Series Standing Order ISBN 978–0–333–71708–0 hardcover
Series Standing Order ISBN 978–0–333–71110–1 paperback

You can receive future titles in this series as they are published by placing a standing order. Please contact your bookseller or, in case of difficulty, write to us at the address below with your name and address, the title of the series and one of the ISBNs quoted above.

Customer Services Department, Macmillan Distribution Ltd, Houndmills, Basingstoke, Hampshire RG21 6XS, England

The Chinese State, Oil and Energy Security

Monique Taylor
Postdoctoral Fellow, Nanyang Technological University, Singapore

© Monique Taylor 2014

All rights reserved. No reproduction, copy or transmission of this publication may be made without written permission.

No portion of this publication may be reproduced, copied or transmitted save with written permission or in accordance with the provisions of the Copyright, Designs and Patents Act 1988, or under the terms of any licence permitting limited copying issued by the Copyright Licensing Agency, Saffron House, 6–10 Kirby Street, London EC1N 8TS.

Any person who does any unauthorized act in relation to this publication may be liable to criminal prosecution and civil claims for damages.

The author has asserted her right to be identified
as the author of this work in accordance with the Copyright, Designs and Patents Act 1988.

First published 2014 by
PALGRAVE MACMILLAN

Palgrave Macmillan in the UK is an imprint of Macmillan Publishers Limited, registered in England, company number 785998, of Houndmills, Basingstoke, Hampshire RG21 6XS.

Palgrave Macmillan in the US is a division of St Martin's Press LLC,
175 Fifth Avenue, New York, NY 10010.

Palgrave Macmillan is the global academic imprint of the above companies and has companies and representatives throughout the world.

Palgrave® and Macmillan® are registered trademarks in the United States, the United Kingdom, Europe and other countries

ISBN: 978-1-137-35054-1

This book is printed on paper suitable for recycling and made from fully managed and sustained forest sources. Logging, pulping and manufacturing processes are expected to conform to the environmental regulations of the country of origin.

A catalogue record for this book is available from the British Library.

A catalog record for this book is available from the Library of Congress.

Transferred to Digital Printing in 2014

*To my parents, Lloyd and Colleen, and
my sister, Tamarind*

Contents

List of Figures and Tables	ix
Acknowledgements	x
List of Abbreviations	xii

1 A Party-State Centred Explanation of Policymaking in China's Oil Sector — 1
- China's oil security dilemma — 4
- China's oil policy approach — 8
- Building oil state capacity — 12
- Political and bureaucratic hierarchies of authority — 14
- Beijing's evolving conceptualisation of energy security — 17
- Socioeconomic dimensions of oil policy — 20
- Oil industry and firm development under market transition — 22
- Research methods — 25
- Organisation and chapter synopses — 26

2 Sectoral Governance and State Capacity — 31
- The concept of state capacity — 34
- State capacity and autonomy — 37
- Contrasting views of state capacity in China — 39

3 The Interplay of Elite and Bureaucratic Power — 50
- The FA perspective on China's energy policymaking — 53
- The neglected role of the CCP — 59
- Strengthening CCP authority and legitimacy through intra-party reform — 61
- A case for bureaucratic authoritarianism — 66

4 The Socialist Era of Oil Self-Sufficiency (1949–1977) — 70
- Early lessons in the pitfalls of foreign oil dependency — 73
- The pursuit of self-reliance during the Great Leap Forward — 76
- The realisation of oil self-sufficiency under Mao — 77
- Construction of the third front and the Cultural Revolution — 81
- The slow shift away from economic autarky in oil sector development — 84

5	**Decentralisation and Corporatisation of the Oil Sector (1978–2002)**	88
	Foreign participation in oil sector development under reform and opening	90
	The creation of China's NOCs: CNOOC, Sinopec and CNPC	92
	The oil price reform conundrum	98
	Bureaucratic reform and decentralisation	105
	China's oil industry in the first Reform Era	107
6	**Rebuilding Oil State Capacity (2003–2013)**	112
	Government restructuring since 1998: Building the regulatory state	118
	Recentralising the oil governance regime	127
	Oil policy formulation and implementation	143
7	**China's National Oil Companies 'Go Global'**	149
	Corporate governance with Chinese characteristics	153
	Ownership and regulation of China's NOCs	156
	NOC leadership	161
	The 'go global' policy	166
	Beijing's oil diplomacy	169
8	**Authoritarian State Capacity in a Liberal World Order**	175
	The rise of China's market authoritarianism model	178
	Implications of China's state-led oil strategies for business and politics	182
Bibliography		188
Index		207

List of Figures and Tables

Figures

1.1	China's Primary Energy Consumption in 2012 (2735.2 million tonnes oil equivalent)	7
1.2	China's Crude Oil Imports by Region in 2010	11

Tables

6.1	Members of the NEC in January 2010	140
6.2	Members of the NEC as of September 2013	141
7.1	Major share ownership of Sinopec, PetroChina and CNOOC	157
7.2	Composition of the Board of Directors in China's NOCs	163

Acknowledgements

I have received much support, helpful advice and encouragement from numerous individuals and organisations throughout the years of research and writing it took to produce *The Chinese State, Oil and Energy Security*. Support for the study, in the form of research funding and academic supervision, came principally from the School of Political Science and International Studies at the University of Queensland, and I am very appreciative of the resources and facilities this world class institution has provided.

I was fortunate to have the opportunity to test my ideas as my project evolved by presenting at the following conferences and workshops, and would like to thank the hosts and participants for their insightful comments: the Oceanic Conference on International Studies in Auckland, July 2010, the Australian Political Science Association Conference in Melbourne, September 2010, the Third International Political Economy Workshop at Griffith University in Brisbane, December 2010, the International Studies Association Annual Conference in Montreal, March 2011, the Emerging Leaders' Dialogue at Peking University, Beijing, July 2011, the ISAS-Griffith University Workshop on The Political Economy of State-Owned Enterprises at the National University of Singapore, August 2011, and the C9-Go8 HDR Forum: Clean Energy and Global Change for the Future Conference at Tsinghua University, October 2011. In researching the book I also benefitted from assistance during fieldwork and conference trips to China, Hong Kong and Singapore. I am thankful to everyone with whom I engaged in formal interviews, through to casual conversations about Chinese politics over coffee. All these experiences helped to shape this book.

There are several individuals who should be thanked by name. I extend my deeply felt gratitude to my former supervisors, Professor Stephen Bell and Associate Professor David Martin Jones, for their valuable insights, guidance and wise advice. One could not have asked for better teachers and mentors. I would also like to thank Dr Jian Yang, who inspired my interest in China during my undergraduate and postgraduate years at the University of Auckland, and encouraged me to continue my graduate studies. I am also grateful to Dr Owain Williams for giving me sound advice on book proposals and a nudge to send off my own one to Palgrave Macmillan. In addition, I would like to acknowledge how wonderfully supportive and patient the editors at Palgrave have been while I worked on my book manuscript.

Finally, I am enormously grateful and much indebted to my parents and sister for their remarkably enduring and unwavering support for my studies and decision to pursue an academic career. Their constant encouragement gave me the confidence to complete the project. This book is for them.

List of Abbreviations

ASEAN	Association of Southeast Asian Nations
BA	Bureaucratic Authoritarianism
BRICS	Brazil, Russia, India, China and South Africa
CASS	Chinese Academy of Social Sciences
CCP	Chinese Communist Party
CDB	Central Development Bank
CEO	Chief Executive Officer
CFO	Chief Financial Officer
COD	Central Organisation Department
CNOOC	China National Offshore Oil Corporation
CNPC	China National Petroleum Corporation
COCOM	Coordinating Committee for Multilateral Export Control
COSTIND	Commission of Science Technology and Industry for National Defense
CPCIA	China Petroleum and Chemical Industry Association
CSRC	China Securities Regulatory Commission
EBL	Energy-Backed Loan
Eximbank	Export-Import Bank of China
FA	Fragmented Authoritarianism
FAWC	Central Work Conference on Foreign Affairs
FDI	Foreign Direct Investment
FOB	Freight on Board
FOCAC	Forum on China-Africa Cooperation
GDP	Gross Domestic Product
IEA	International Energy Agency
IMF	International Monetary Fund
INOC	International National Oil Company
IOC	International Oil Company
IPO	Initial Public Offering
IR	International Relations
GFC	Global Financial Crisis
LLC	Limited Liability Company
MFA	Ministry of Foreign Affairs
MII	Ministry for Industry and Information
MOE	Ministry of Energy
MOF	Ministry of Finance

List of Abbreviations xiii

MOFCOM	Ministry of Commerce
MOFERT	Ministry of Foreign Economic Relations and Trade
MOFTEC	Ministry of Foreign Trade and Economic Cooperation
MOG	Ministry of Geology
MPI	Ministry of Petroleum Industry
NATO	North Atlantic Treaty Organization
NDRC	National Development and Reform Commission
NEA	National Energy Administration
NEC	National Energy Commission
NELG	National Energy Leading Group
NOC	National Oil Company
NPC	National People's Congress
NYMEX	New York Mercantile Exchange
NYSE	New York Stock Exchange
ODI	Overseas Direct Investment
OECD	Organisation for Economic Cooperation and Development
OFDI	Outward Foreign Direct Investment
ONELG	Office of the National Energy Leading Group
OPEC	Organization of Petroleum Exporting Countries
PBC	People's Bank of China
PLA	People's Liberation Army
PRC	People's Republic of China
PSA	Production Sharing Agreement
SASAC	State Asset Supervision and Administration Commission
SCO	Shanghai Cooperation Organisation
SDPC	State Development Planning Commission
SEC	Securities and Exchange Commission
SEO	State Energy Office
SETC	State Economic and Trade Commission
Sinochem	China National Chemicals Import and Export Corporation
Sinopec	China Petrochemical Corporation
SOA	State Oceanic Administration
SOE	State-Owned Enterprise
SPC	State Planning Commission
SPCIB	State Petroleum and Chemical Industry Bureau
SDPC	State Development Planning Commission
SPR	Strategic Petroleum Reserve
UNDP	United Nations Development Programme

USCC	United States-China Economic and Security Review Commission
WEPP	West-East Pipeline Project
WTO	World Trade Organization

1
A Party-State Centred Explanation of Policymaking in China's Oil Sector

This volume traces the development of China's state-led oil strategy, which is frequently referred to in the extant literature as 'neomercantilist' in orientation. This statist strategy is in contrast with a liberal market-led approach to energy security, which relies on markets to allocate oil resources and prescribes only a minimalist facilitator role for the state. China's state-led oil policies entail the use of top-down party-state authority and control in order to undertake domestic and international oil production for the purpose of ensuring a secure, stable and affordable oil supply. The central party-state's institutional capacities and policy instruments, such as national oil companies (NOCs) and central planning agencies, enable Beijing to pursue this statist approach. In explaining China's oil policy rationale and implementation, this study provides a historical analysis of party-state institutions and the Chinese policy process. In doing so, it shows that the central party-state has, in recent years, expanded and strengthened the political, organisational and fiscal capacities that permit it to exert centralised, top-down authority, while at the same time retaining the incentives and dynamism that were created through decentralisation of the market-oriented players during the earlier stages of oil industry development. This party-state centred explanation of institutional change relies on historical narrative to show how state capacities in the oil sector have been accumulated, built and improved over time. Where institutional change within this sector has occurred, usually at critical junctures in China's economic development and market transition, the Chinese Communist Party (CCP) elites have driven the process.

This argument runs counter to conventional accounts of China's oil strategy, which are typically informed by an institutional approach to public policy analysis called fragmented authoritarianism (FA). Developed by Lieberthal and Oksenberg in the late 1980s, the FA model privileges the role of mid-level bureaucratic organisations in shaping policy outcomes in China. It broadly claims that the decentralised decision-making authority and fragmented structure that allegedly characterises the Chinese political system produces competition, conflict and extensive bargaining among various political actors, a dynamic that thwarts effective policy implementation (Lieberthal and Oksenberg 1988). The key findings of studies that rely on this particular theoretical understanding of the Chinese policy process as being disjointed and bogged down, tend to reveal a dysfunctional government and bureaucracy largely incapable of producing coordinated, coherent and effective energy policies (see for instance, Kong 2005, 2006 and 2010; Zha 2006; Downs 2004a, 2006 and 2008a; Lester and Steinfeld 2006 and 2007; Meidan et al. 2009; Yeo 2009a). According to this FA view China's oil strategy is not state-led, but is instead the product of bottom or middle-up initiatives driven primarily by the NOCs (Downs 2006 and 2008a; Houser 2008; Jiang and Sinton 2011). The FA model was useful in explaining a particular period of policy-making in China, namely the decade of liberalisation and decentralisation that emerged with the onset of the Reform Era ('reform and opening', *gaige kaifang*) in 1978. This decade was characterised by a marked shift within the party-state from central command and control as well as Mao's dictatorial rule, to more "horizontal interorganisational bargaining" (Bell and Feng 2013: 113).

While it remains useful in explaining bureaucratic authority in the post-Mao era, the FA model provides an incomplete understanding of policy dynamics and fails to account for the substantial change that China's institutional landscape has undergone since the 1980s. FA also lends itself to the inaccurate, yet popular, view of China muddling through the Reform Era, ignoring the decisive, purposeful and reasonably successful actions of party-state leaders that have driven the reform process. This decisiveness has been especially evident since the early 1990s when significant recentralising and self-strengthening efforts were undertaken by Beijing in the aftermath of the Tiananmen protests and collapse of the Soviet Union and communist party-states in the Eastern Bloc (Yang 2004; Shambaugh 2008). The focus of the past decade in particular has been on building capacity in the 'strategic sectors' of China's economy, such as the oil industry. The party leadership considers these strategic sectors vital to China's economic growth and development, and social stability, and as such they were never intended to 'grow out of the plan' and eventually privatise, in contrast to the country's non-state sectors (Naughton 1996).

Arguably now, more than ever, the central party-state is equipped to wield top-down authority and solicit compliance from various bureaucracies through a variety of CCP controls and career incentive structures within the nomenklatura system, which has been tightened in recent years and allows the party to determine senior personnel appointments throughout the state sector, and also through various monitoring mechanisms. The central government's control over the banking system and oil pricing, as well as ownership of the NOCs, accounts for the significant power of the party-state to set policy for the oil industry. These centrally controlled political, organisational and financial instruments effectively counter the worst effects of bureaucratic fragmentation, hence mitigating some of the institutional shortcomings emphasised by the FA model. This is especially the case when a particular economic sector commands attention from the top leadership, which is what happened to the oil industry from roughly 2003 onwards. The centralised and pervasive power of the CCP is a particularly significant yet largely neglected variable, and is one that is increasingly advanced by a handful of scholars such as Naughton (2008a), Bell and Feng (2009 and 2013), Pearson (2007) and Shambaugh (2008) as being fundamental to understanding policymaking in China. In this study 'bureaucratic authoritarianism' (BA) is advanced as a more appropriate model since it provides an elite-driven account of institutional change and top-down policymaking by emphasising "the rise of bargaining within a *hierarchical* state system" (Bell and Feng 2013: 115).

At various stages of China's economic development the CCP leadership has clearly demonstrated its capacity to reach in, reorganise and restructure the Chinese state. Since the FA model provides a static description of how the state apparatus functions, it struggles to explain how institutional change has occurred within the party-state. Furthermore, in bracketing the influence of elite power, this model certainly does not perceive the significance of the CCP in effecting institutional change. While political authority in the oil sector is primarily a top-down phenomenon, the relationship between the central party-state and the NOCs is reciprocal in the sense that the oil companies are granted space to advise policymakers by virtue of the fact that they are repositories of specialised knowledge and expertise, and possess strong informal connections to party-state leaders. This dynamic has led Kong (2010: 27–28) to suggest that China's oil policy is shaped by the central government and the NOCs' 'co-governance' of the oil industry, which implies that both actors have equal decision-making input. However, a close examination of the interplay between elite and bureaucratic power within the party-state shows that the Chinese leadership remains the pivotal and decisive player that ultimately chooses policy content and determines strategic direction for the oil industry. As such it is the steep hierarchy of authority within the central party-state that governs this collaboration between the Chinese government and the NOCs.

Central party-state capacities were strengthened and have expanded since the early 1990s, resulting in the emergence of a clearer energy security agenda and blueprint for economic and industrial development within the strategic sectors of China's economy. The capacity of the party leadership to drive reform and institutional change is the central dynamic shaping oil industry development. This study explores governance arrangements within this strategic sector, how they impact energy policy articulation and implementation, and ultimately how this is reflected in China's domestic and international energy behaviour with respect to both upstream and downstream oil activities. Analysing these various facets of policymaking in China's oil sector involves exploration of the policy orientation and central directives espoused by the leadership of the CCP (who determine the overarching framework of China's energy policy, laid out explicitly in the government's Five-Year Plans and other policy documents), and how these have been implemented by key party-state bureaucracies including the State Council, the National Development and Reform Commission (NDRC), the newly formed National Energy Commission (NEC) and its standing body – the National Energy Administration (NEA), the Ministry of Foreign Affairs (MFA), the Ministry of Finance (MOF), the Export-Import Bank of China (China Eximbank), the China Development Bank (CDB), the State Asset Supervision and Administration Commission (SASAC) and, most importantly, the NOCs – China National Petroleum Corporation (CNPC), China Petrochemical Corporation (Sinopec Group), China National Offshore Oil Corporation (CNOOC) and Sinochem Group (Sinochem).

China's oil security dilemma

China's role as a major driver of world oil demand growth over the past decade has generated apprehension, even alarm, among other oil consuming countries, notably the United States and Japan. Here the concern does not simply revolve around China's ever-increasing oil requirements to fuel its rapidly expanding economy, nor its newly-acquired status as the world's largest energy consumer and second largest oil consumer and importer behind the United States (despite the fact that it is the fourth largest oil producer) (*People's Daily* 2010b; IEA 2007: 265; IEA 2010), but also the perceived manner in which the country seeks to achieve energy security, through both its domestic and international oil activities. For instance, whilst the official rhetoric espoused by the Chinese leadership favours increased marketisation, China's oil sector remains under tight state control and entails a regime of price controls and subsidies. This encourages the inefficient allocation and excessive consumption of oil, thus maintaining domestic demand at artificially high levels when world oil prices are high, which was evident during the oil shock of 2007–2008. On the international stage China's preference for long-term bilateral oil supply contracts (termed equity oil) with oil-producing countries, which allegedly 'lock up' foreign oil supplies, may be considered a threat to the effective functioning of world oil markets, and the competitiveness of other NOCs and international oil companies (IOCs), as well as to the energy security of other oil importers (Pei 2006a). Other aspects of China's statist oil strategy, including the extensive state-backed financial support and oil diplomacy used to augment the overseas investments of Chinese NOCs have also raised concerns among IOCs and oil consuming countries. Given current projections of China's oil demand, worries surrounding its potential impact on oil affordability and accessibility may intensify over the coming decades. The IEA forecasts that China's oil demand will account for the largest absolute increase in global oil demand growth through to 2035. It estimates that China will experience an average annual increase in oil demand of 2.2 per cent, with its consumption increasing from 9.0 mb/d in 2011 to 15.1 mb/d by 2035 (IEA 2012: 86). This increase would account for half the net increase in oil demand worldwide (IEA 2012: 86).

Dependency on imported crude oil accounted for 53.8 per cent of China's total oil demand in 2010 (IEA 2012: 6). The 2010 edition of the IEA's *World Energy Outlook* predicts that this oil import dependence is set to rise to 84 per cent by 2035. Within China's energy mix oil remains less important than coal, comprising 18 per cent of total primary energy demand in 2012 (see Figure 1.1 for the composition of China's energy consumption by fuel) (BP 2013). Demand for oil is the fastest growing component of China's total energy demand, and poses the greatest energy security challenge for the country due to growing import dependency, world oil price volatility, strategic chokepoints in

China's oil supply routes such as the Strait of Malacca, political volatility in many oil-exporting countries and international competition for increasingly scarce oil resources. China's domestic oil production has stagnated, especially in onshore crude oil-producing areas (most of these oil fields have already reached or passed their peak), and the ever-widening gap between China's oil production and oil consumption must now be filled with imported oil. China's NOCs, as latecomers to the global oil industry, have faced a steep learning curve and experienced some major setbacks in their pursuit of overseas mergers and acquisitions, a prime example being CNOOC's thwarted attempt to purchase California-based Unocal in 2005. Hence China's sharply increasing oil import dependency now occupies a central position on Beijing's policy agenda, with Zheng Bijian (a lead thinker and advisor to the Chinese leadership) considering energy security to be the top policy challenge facing China (Chen 2011: 600). Energy policy is also intimately connected to social, economic, foreign and security policy, and has altered Beijing's calculations of the national interest, as concerns about energy security, particularly conceived in terms of the dilemma of growing oil import dependency, appears to be an important factor that influences China's perception of its external security environment and drives foreign policy (Zweig and Jianhai 2005; Ziegler 2006).

China's demand for oil has been dictated by the nature of its economic development path and changing industrial profile. During the first twenty years of economic reform, from 1980 to 2000, economic growth in China was primarily dependent on light industry (mainly low-end manufacturing) and small-scale private enterprise. These industrial sectors are labour, rather than energy intensive. Hence while economic growth in this period occurred rapidly, energy efficiency gains were also apparent. From the early 2000s onwards, the Chinese economy's energy intensity began to increase significantly. Breakneck economic growth, entry into the World Trade Organization (WTO) in 2001 (causing rapid growth in China's trade volume, accelerating energy consumption), increasing urbanisation, motorisation and structural changes within China's economy led to increases in China's oil demand that far exceeded earlier predictions of energy demand growth. Indeed the boom in economic growth and the surge in the output of heavy industry from 2002 to 2005 in particular caught outside observers such as the IEA by surprise (Andrews-Speed 2011: 15–16). At this time the Chinese leadership shifted structural economic

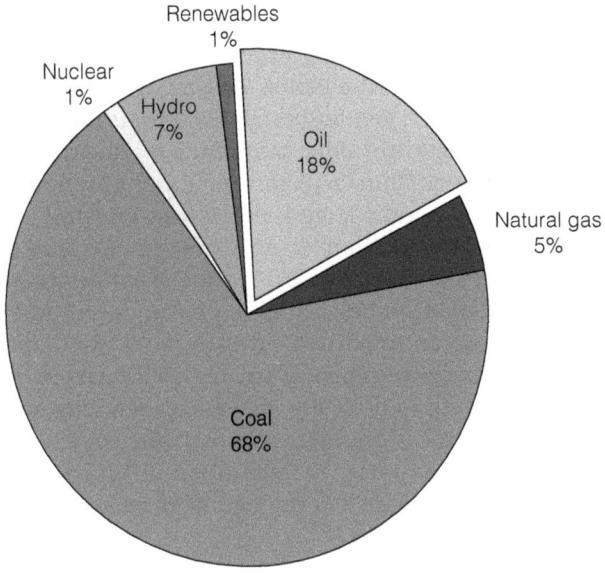

Figure 1.1 China's Primary Energy Consumption in 2012 (2735.2 million tonnes oil equivalent)
Source: BP Statistical Review of World Energy June 2013

emphasis at the national level away from light industry and back toward heavy industry, which is highly energy intensive (Kahn and Yardley 2007). The reasons behind the shift towards heavy industry also dates back to 1997 when China's leaders were concerned that the Chinese economy might suffer the same fate as other East Asian economies and enter recession in the wake of the Asian Financial Crisis. In response Beijing put together an economic stimulus package, which provided "generous state financing and tax incentives to support industrialisation on a grand scale" (Kahn and Yardley 2007). It worked remarkably well. Kahn and Yardley (2007) note that in 1996 China and the United States each accounted for 13 per cent of global steel production, but by 2005 the United States' share had dropped back to 8 per cent whilst China's had risen to 35 per cent. China also manufactures around half of the world's cement and glass and about a third of its aluminium (Kahn and Yardley 2007).

Urbanisation in particular has caused heavy industry to grow as it creates demand for steel, cement and other industrial materials used to expand critical infrastructure and housing. In addition to the development of these energy intensive industries, a burgeoning consumer class and rapidly expanding civil aviation and car fleets further increase China's appetite for oil. The transportation sector is now the principal driver of oil demand growth, alone accounting for one-third of China's total oil consumption. China's phenomenal motorisation boom has seen it become the world's largest and fastest growing car market. Attempts to slow the addition of new cars on the road would be difficult at this stage of China's modernisation as car ownership is symbolic and aspirational among Chinese consumers (Andrews-Speed 2011: 1), and is also an important "growth point" in domestic consumption, which the government is trying to stimulate (Cheng 2008: 310). Instead of trying to limit the increase of new cars on the road, Chinese authorities focus on improving fuel efficiency (Cheng 2008: 310). Hence the contemporary manifestation of China's economy is highly energy intensive, with an ever-increasing reliance on oil.

China's oil policy approach

Since the early 1990s China has undertaken a process of strategic adjustment in response to two major energy security challenges. The first challenge was growing oil import dependency, which began with the country's shift from net oil exporter to net oil importer in 1993, ending thirty years of oil self-sufficiency. This move away from self-sufficiency also came at a time of rapid modernisation whereby the Chinese leadership was under tremendous pressure to deliver strong economic growth, thus exacerbating the perceived insecurity and vulnerability derived from foreign oil dependency (Yergin 2006: 77). The second challenge emerged a decade later with the hike in world oil prices beginning in 2003 and culminating in the oil shock of 2007–2008. In five years crude oil prices increased nearly 500 per cent, from roughly US$30 in 2003 to a record price of US$147 in July 2008. During this time the Chinese economy was also undergoing structural adjustment emphasising the development of heavy industries, which are of course energy-intensive, particularly in terms of fossil fuel consumption. This further heightened China's energy crisis and catapulted oil security to the realm of high politics. Initially oil import dependency confronted the Chinese leadership with a new range of national security and geopolitical considerations, arising from the need to secure foreign oil supplies. However, in combination with inexorably rising oil prices, these concerns soon gave way to other systemic economic, industrial, social and political problems within the Chinese state, namely those relating to social equity and stability, industrial organisation and competitiveness and the need to maintain high economic growth rates. The deliverance of strong economic growth is of course the touchstone of CCP legitimacy and long-term survival, and secure, reliable and affordable sources of energy are fundamental to achieve this.

Faced with these energy security dilemmas Chinese authorities had several policy options from which to choose. Broadly, oil-importing countries can pursue either economic nationalist or liberal market approaches to the problem of energy security. Economic nationalists or neomercantilists, favour the concept of energy independence and define energy security in terms of the security of energy supply, which they seek to ensure through state-led energy production and distribution both at home and abroad (if self-sufficiency is unattainable), with foreign oil assets often secured through government-to-government contracts. Liberal market approaches, on the other hand, emphasise energy interdependence where global energy needs are satisfied via free market competition among domestic and international energy consumers. Here the market instrument requires little, if any, state involvement in the energy commodity chain. Ikenberry (1986: 113–116) identified a third approach of competitive accelerated adjustment, which targets the demand side of the energy equation through industrial competitiveness, for instance, by encouraging industrial energy efficiency and conservation with the aim of reducing national fossil fuel consumption. These three categories capture broad policy emphasis and approach. However it is important to recognise that energy policy is inherently complex and multifaceted and usually incorporates a variety of strategies in practice. China's response to a range of exogenous and endogenous factors such as oil resource constraints, international oil price volatility and instability in oil-producing countries, broadly conforms to a state-led strategy (Kong 2005: 56; Zweig and Jianhai 2005; Zhao 2008: 207; Erickson and Collins 2007: 683), as Kong (2005: 56) neatly states: "Distrustful of the market, and suspicious of other major energy players in the international market, the Chinese leadership relies on the state-centred approach, or economic nationalism, rather than a market approach to its energy security." That being said, China also pursues the other approaches to varying degrees, and it is important to note that while Beijing shows a clear preference for the state-led approach, the country remains reliant on international oil markets to satisfy the bulk of its oil needs.

Beijing's strategic focus on the security of oil supply manifests itself in a range of state-led policy approaches that are executed by various government agencies and the NOCs both at home and abroad. One of China's main strategies has been to step up domestic oil production to try and counteract the decline in domestic oil output, thus reducing oil import dependency. The cost of domestic oil production in China provides further evidence of the pursuit of a strategic, as opposed to market approach to energy security, as China's NOCs have invested heavily in trying to develop methods of extracting extra barrels of crude oil and "pump more crude at almost any cost" (Graham-Harrison 2008). Mike Wittner, global head of research at Société Générale, claims that this approach is the result of oil being a "national security" issue in China, whereas "... in most countries in the world it's about investment and production costs, is it worthwhile making an investment?" (Graham-Harrison 2008). Nonetheless, these developments are unlikely to significantly offset China's future demand for oil imports. In light of the insufficiency of China's domestic oil supply, the pursuit of equity oil abroad has been another prominent strategy, as Pei (2006a) states, "Beijing's favourite method appears to be getting direct equity stakes in oil fields." This approach has caused concern among other oil-importing countries, and particularly among Washington policymakers, due to the popular perception that China is "locking down" oil supply, and overpaying for oil resources (indicating that profit-oriented considerations are not solely driving the NOCs' investments). The NOCs spearhead China's oil strategy, receiving strong financial backing and guidance on where to invest from the Chinese government. Ownership of foreign oil fields is perceived by Beijing to improve energy security, as "the Chinese leadership believes that China cannot depend on Western oil companies or the international oil market in times of crisis" (Cheng 2008: 314).

Arguably the most important strategy for improving China's energy security has been to geographically diversify the sources of foreign oil supply (Pei 2006a). Beijing's focus on diversification became particularly prominent after the Iraq War in 2003. China is primarily reliant on Middle Eastern oil and aims to decrease dependence on this region due to its political instability and geostrategic domination by the United States. A diversification strategy minimises the potential for oil supply disruption because oil is sourced from multiple suppliers from different regions of the world. To achieve diversification Beijing has relied on 'state direction' over the NOCs, the pursuit of equity stakes in foreign oil supply (part of the 'go global' or 'going out' strategy, *zou chuqu*), and oil diplomacy (Vivoda and Manicom 2011: 228). China has made strong inroads into Africa, Latin America and Central Asia in its pursuit of alternative sources of oil supply within a relatively short period of time (see Figure 1.2 for China's crude oil imports by region). Relatedly, another strategy has been to integrate energy security objectives with foreign policy and diplomatic efforts. This has resulted in a significant increase in the number of bilateral relationships between China and oil-rich countries, and the number of oil deals brokered by Beijing. Controversially China has pursued a significant number of oil deals with 'pariah' states such as Sudan, Iran, Libya (under Qaddafi) and Myanmar. Furthermore, China has also sought to secure and diversify its oil supply routes. This concern has added impetus to the development of a blue water navy (as most of China's oil is delivered by sea), and oil pipelines into China from parts of Russia, Central Asia and Myanmar. Lastly, development of a strategic petroleum reserve (SPR) has been underway since 2004, and by 2010 held roughly thirty days of imported oil.

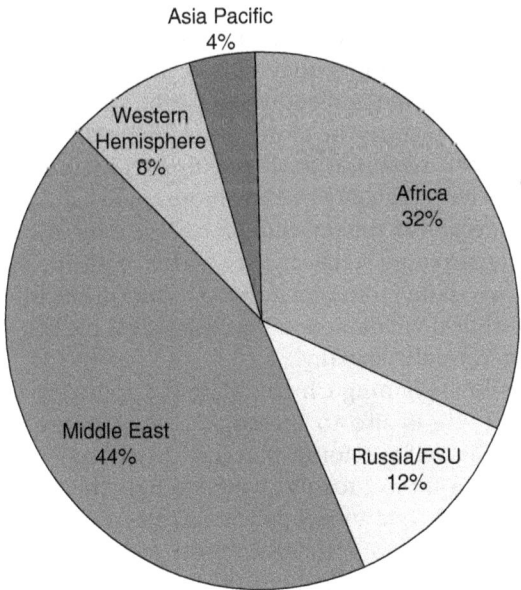

Figure 1.2 China's Crude Oil Imports by Region in 2010
Source: IEA Information Paper (Jiang and Sinton 2011)

Building oil state capacity

Various attempts have been made to explain why China pursues a predominantly statist oil policy approach, the more nuanced of which focus on oil state capacity. Here 'state capacity' is defined simply as the ability of the state to formulate and implement policy. Ultimately Beijing pursues state-led oil exploration, production and distribution as it considers this an effective strategy to safeguard economic growth and development, and social stability at this particular stage of China's economic development and market transition. The oil industry also plays an important role in China's economy as a major source of employment – CNPC alone employs 1.6 million people (Alexander 2012). Hence for a variety of socioeconomic, industrial, political and security reasons China views oil as a strategic good, rather than a regular commodity (Andrews-Speed et al. 1999: 2). While market-oriented reform has been a chief feature of the Reform Era, it has not been extended to the oil industry because the Chinese leadership believes that this sector is currently "far too important to be left to market forces alone" (Kreft 2006). While a range of domestic factors motivate the preference for economic nationalism in the oil sector, equally as important is the availability and continual improvement of the institutions and policy instruments that enable the central party-state to pursue this energy policy option. In other words, the party leadership has perceived and addressed China's oil industry development and energy security in terms of an increasingly sophisticated range of political, organisational and financial capacities. China's energy policies have arguably become more coordinated and effective over the past decade as the result of an overall tightening of strategic focus and strengthening state capacity within both the oil sector and the central party-state more generally. As mentioned earlier, this is an elite-driven political phenomenon that challenges the conventional FA account of energy policymaking.

In addition to explaining China's state-led approach to energy security, the purpose is also to investigate whether the country's oil sector reforms, as well as reforms that have been implemented within the Chinese state more broadly, have strengthened this industry's ability to deliver effective energy policy outcomes. The methodology is to compare three phases with distinct oil policy regimes: a largely autarkic phase running from 1949 to 1977, a phase of limited opening, and the decentralisation and corporatisation of market-oriented players from 1978 to 2002; and a more coordinated phase (the result of efforts to build state capacity in China's oil sector) running from 2003 through to 2013. The evidence shows that during the earlier phases of oil industry development, bureaucratic organisation was predominantly characterised by FA. However, over the past decade the central party-state has instituted reforms aimed at 'defragmenting' the state apparatus in the sense of creating "a hierarchical structure of bureaucratic authority"; a strategy also used in the wind energy sector (Lema and Ruby 2007: 3888), though in the case of wind power with arguably less success. The capacities resulting from an enhanced strategic focus on oil coupled with institutional re-organisation within the Chinese state are considered the key determinants impacting policy articulation and implementation, and the overall effectiveness of oil policy outcomes. The focus of this volume looks beyond the effects of mid-level bureaucratic institutions to examine the influence of other institutions and policy instruments, notably those developed by the CCP to solicit compliance from bureaucracies including the NOCs. These are found to be crucial in enabling the party leadership to push through its policy agendas. In terms of policy approach, where the oil industry exhibited a more decentralised and fragmented model of decision-making, a narrow focus on the security of oil supply ensued. By contrast, when organisational structure in the oil sector and the broader party-state began to defragment and centralise, and the top leadership focused its policy efforts on energy security, more coordinated and comprehensive energy policies emerged targeting both demand and supply side management.

China's industrial policy toward the pillar or lifeline industries – the commanding heights of the national economy, namely infrastructure, telecommunications, banking, energy and natural resources – is also central to understanding how oil policy has taken shape. Economic governance of strategic assets in China continues to "work in the service of the party-state rather than the market" (Pearson 2007: 724). To this end the state sector responds to the political imperatives of the party-state, and not solely to economic signals. The state sector of the Chinese economy has been restructured through a series of reforms since 1996, known as 'grasp the large, release the small' (*zhua da fang xiao*). The aim of this reform drive was to downsize the state sector, professionalise state enterprise management, stabilise and unify government ownership (through the establishment of an ownership agency known as the State Asset Supervision and Administration Commission (SASAC)) and improve the efficiency and productivity of the remaining large state-owned enterprises (SOEs) without resorting to full blown marketisation (Naughton 2008c: 147). This handful of strategic industries are characterised by oligopolistic market structures usually comprised of a few large state-owned 'national champions'. The South Korean model influenced the leadership's plan to reorganise the state sector, beginning in the 1990s; a limited number of large state enterprises in several market-protected strategic sectors would come to dominate the Chinese economy, in much the same way as the Korean economy is dominated by the *chaebol* (large, integrated conglomerates with close ties to the government) (Fewsmith 2008a: 210; Dickson 2011: 42). Supervision and control of the remaining SOEs was centralised under the State Economic and Trade Commission (SETC), which later became SASAC, eliminating a "multiplicity of bureaucratic interests" (Fewsmith 2008a: 210). Dramatic improvement in the performance of the state-run sector via direct party-state intervention is another positive indicator of growing state capacity, as Chinese authorities were able to create internationally competitive state enterprises in a relatively short period of time. China's oil sector was one of the key industrial sectors to undergo extensive restructuring, out of which emerged vertically integrated state-run oil and gas holding companies; CNPC, Sinopec, CNOOC and Sinochem (although Sinochem was 'conspicuously excluded' from the 'radical restructuring' undertaken by the State Council in 1998), which have spearheaded China's quest for oil resources "in the name of serving national energy security", both at home and abroad (Chen 2011: 601; Zhang 2003: 185). An unintended consequence of the 'grasp the large, release the small' policy was that the remaining large SOEs amassed even more political leverage to defend their own interests, and in some cases block further reform. In the case of the oil industry, the NOCs are a power base in the political system and a source of political patronage, a problem which is currently being addressed by President Xi Jinping's concerted effort in 2013 to remove the "stalwarts" of NOC influence, notably Jiang Jiemin (former Sinopec chairman and director of SASAC) and Zhou Yongkang (head of internal security and former CNPC general manager) on charges of corruption (*China Economic Review* 2013).

Political and bureaucratic hierarchies of authority

Since it first appeared in 1988, the dominant framework used to analyse policymaking in post-Mao China has been the FA model. Lieberthal and Oksenberg (1988) developed this model in order to explain economic decision-making within what they perceived to be an increasingly fragmented government and bureaucracy, and Lampton (1987) essentially shared this view of the policy process. Lieberthal and Oksenberg (1988) use this model specifically to explain the policy process in China's energy sector during the 1980s. According to the FA model the fragmented structure of government and bureaucracy in China is the central explanatory variable, and one that permeates both policy formulation and implementation. This model focused on formal mid-level government organisations engaging in intensive and extensive horizontal interorganisational bargaining, thus emphasising the central importance of bureaucratic structure to the policy process (Zhao 1995: 233). The principal findings of FA research revealed a "fragmented bureaucratic structure of authority, decision making in which consensus building is central, and a policy process that is protracted, disjointed and incremental" (Lieberthal and Oksenberg 1988: 22). This model claims that the Chinese political system is generally uncoordinated and incoherent, and that these dysfunctional aspects of the bureaucracy shape Chinese politics and influence policy decisions. Since policymaking is shaped primarily by bureaucratic politics according to FA, the flow of political authority is considered to be bottom- or middle-up, rather than top-down. As a result of the allegedly fragmented and decentralised nature of the political system, elite policy agendas could be altered and distorted at the implementation stage by the demands of this fragmented and protracted policy process (Lieberthal and Oksenberg 1988: 22). Despite significant and far-reaching changes to China's economic, administrative and political environment since the 1980s, the FA model continues to inform the majority of Chinese policy and governance studies. The oil industry literature is no exception; scholars such as Kong (2006 and 2010), Downs (2004a and 2006), Lester and Steinfeld (2006 and 2007) and Meidan et al. (2009) argue that the country's energy institutions are largely ineffective in terms of formulating and implementing energy policies.

In contrast to these conventional explanations, this study argues that the FA model is becoming somewhat outdated and needs to be modified to take account of the extensive capacity and institution-building efforts that have occurred since the restructuring of elite leadership in 1993, during what Naughton (2008a: 100) refers to as the 'second era of reform' (1993–present). According to Naughton (2008a: 126) this period of reform is characterised by "recentralisation of resources and a tendency toward a more rule-bound and predictable system," where decision-making became much more decisive. By contrast the 'first Reform Era' (1978–1992) was characterised by a 'decentralisation of power and sharing of resources', whereby policymaking was more incremental and fragmented (Naughton 2008a: 101–102). During the post-Tiananmen period Beijing has sought to recentralise, concentrate and strengthen its political authority, which has resulted in efforts to strengthen political levers of control and defragment government administration and bureaucracy. This trend has been particularly marked towards institutional arrangements within the strategic sectors of the economy, where the emphasis shifted from micro- to macro-level, that is, "the state is no longer directly involved in implementation but instead prefers to provide strategic guidance" (Brødsgaard and Zheng 2004: 12). That being said, central planning agencies, notably the NDRC, continue to use "powerful levers of micro-management, such as investment endorsement" (Yeo 2009a: 742). A historical analysis is useful to reveal how these state capacities have changed over time, with a focus on the causes and drivers of institutional change within the Chinese political system. In addition to pre-dating these more recent reform efforts, the FA model can also be faulted for largely ignoring the role of CCP institutions in the policy process. Already mentioned, the party constitutes another layer of governance within the Chinese political system that is particularly powerful and pervasive. Hence it is not just bureaucratic politics that primarily influence policymaking as suggested by the FA model. The party's levers of influence and control must also be factored into consideration.

Throughout the 2000s the upper echelons of the CCP aimed to strengthen party discipline and loyalty, for example, through an intense rectification campaign (Shambaugh 2008: 129–130). Consolidating the ruling status of the party has been a priority for the Politburo and clearly influences policy outcomes. A more appropriate model that explains the energy policy process is that of BA, which swings the pendulum firmly back towards an elitist account of the Chinese political system, emphasising the steep power gradient that exists within the party-state (Hamrin and Zhao 1995; Bell and Feng 2009 and 2013). Advancing such a claim does not lend itself to the inaccurate view that the Chinese state is a unitary or monolithic entity, but rather suggests that Beijing has improved and expanded various state capacities that permit it to exert effective top-down political authority. This power is evident in the party leadership's ability to reach into the Chinese state and reconfigure institutional arrangements. Another significant example is the Chinese leadership's capacity to launch far-reaching anti-corruption campaigns, charging high-level leaders with corruption, the underlying purpose of which is to eliminate or purge an opposition faction or powerful vested interest group within the power structure. Currently Xi Jinping is using anti-corruption crackdowns to reconfigure power within the Chinese state and tackle the powerful SOEs and their allies that are hindering further state sector reform. It is important to note that the extent to which the central party-state is able to wield this top-down authority is variable across policy sectors. The findings in this study should not be extrapolated to the Chinese economy writ large, especially since the oil sector is, in many ways, atypical. For instance, many policy sectors in China are becoming pluralistic with the arrival of new socio-political players, which complicates the decision-making process. This is not the case for the oil sector, and many other state sectors, as all the major players are located within the party-state sphere. Furthermore, even relative to other energy sectors the oil industry possesses a more simplified market structure, with only three main NOCs engaging in limited and managed domestic competition.

Beijing's evolving conceptualisation of energy security

The core concern of energy policy is energy security, which is broadly referred to as 'the uninterrupted availability of acceptable energy sources at affordable prices' in order to support the functioning of an economy and social development (IEA website). This comprehensive definition captures the geological (availability), geopolitical (accessibility), economic (affordability), and environmental and social dimensions (acceptability) of energy security (Jewell 2011: 9). In terms of how these various dimensions of energy security are weighted, most definitions tend to emphasise the physical availability and economic cost of energy supplies (Jewell 2011: 9). Energy security is a context-dependent concept, and the specific nature of energy security problems and their proposed solutions will vary widely. Up until recently China's conceptualisation of energy security was focused almost exclusively on supply-side oil security. In this regard Yergin et al. (1998: 36) went so far as to say that "many describe Beijing's policy options in ways that come perilously close to the shortage-equals-security-threat scenarios of the 1970s". Andrews-Speed et al. (2002) suggest that China's approach to energy security is strategic and, as such, relies heavily on political as well as economic methods. Especially since the early 2000s when China's energy demand began to soar, security of oil supply has become highly politicised. Since China enjoyed oil self-sufficiency for thirty years, up until 1993, its traditional definition of energy security tended to revolve around notions of self-sufficiency and self-reliance, which were both reinforced by a pre-reform political culture stressing these principles in relation to the entire Chinese economy.

The rapid and unexpected shift to net oil import dependency essentially saw the extension of this traditional preference for self-sufficiency into the international arena. For example, China's foreign equity investments represent an attempt to own and control oil in the ground, even if it is located in another country. The various dimensions of China's growing appetite for oil, and dependence on foreign oil supply, is neatly summarised by Yergin (2011: 208): "These are the new commercial realities – China as a growing consumer of oil, China as an increasingly important participant in the world oil industry. But there is also a security dimension, which arises from growing dependence for a country for which 'self-reliance' had been such a strong imperative for so many years." Up until the early 2000s the problem of energy security was not yet a top policy priority. From around 2003 this began to change as an unanticipated jump in Chinese oil demand, rising international oil prices and a series of domestic energy crises saw the leadership turn its attention towards the growing risks to energy security (Yergin 2011: 210). As a result Beijing's narrow focus on the security of oil supply and concerns about external dependence began to broaden and include other sources of energy insecurity such as "China's unreliable, inefficient and heavily polluting energy system" (Kennedy 2010: 143). China's energy institutions in particular came to be viewed as inadequate and unable to formulate and implement effective energy policies (Kong 2005 and 2006; Downs 2006; Lester and Steinfeld 2006 and 2007). In 2003 the Hu-Wen leadership rolled out the concept of 'conservation-minded society' (*jieyue shehui*), the energy and resources component of the Scientific Development Concept (*kexue fazhanguan*), indicating that the environmental dimension of China's energy security has formally entered the energy policy agenda (Constantin 2007: 6–7; Lam 2006: 40–44).

In terms of Beijing's political rhetoric, Kennedy (2010: 146) detects a shift away from a narrowly conceived conception of energy security equating solely to security of oil supply, to an expanded concept that addresses demand-side management. For instance, in 2003 former Chinese President Hu Jintao espoused a traditional notion of energy security by expressing concern over the security of China's oil imports

(Kennedy 2010: 146). However, in 2006 Hu formally discussed a 'new energy security concept' at the G8 Summit; a concept that focuses on reducing domestic energy demand, pursuing research and development to improve energy technologies and encouraging international cooperation to increase oil supply (*China-Gov.cn* 2006). Policy practice has matched this rhetoric as the former Hu-Wen government set ambitious targets for reducing energy intensity and upgrading energy institutions and infrastructure, which have been largely successful, though not entirely unproblematic in their execution. Hence energy policy in China has shown a more recent interest in tackling the demand-side of the energy security equation, for instance, by targeting energy efficiency and conservation, and restructuring the energy mix by substituting oil with renewable energy sources where possible (Cheng 2008: 302). These objectives dovetail with China's strategic approach to energy security in relation to oil as well, since they serve to reduce oil dependency and are more consistent with an overall preference for greater energy independence and the development of robust energy systems. Nevertheless, the problem of energy security is still defined in strategic terms, and while energy policies have become more comprehensive and nuanced, security of oil supply remains a core concern due to the vulnerability that arises from foreign oil dependency. The continuing importance of energy security has extended to the new fifth generation Chinese leadership, as Xi Jinping's first foreign visit was to Moscow, the main purpose of which was to conclude a major natural gas supply pact as well as a raft of other energy agreements including the doubling of Russian oil supplies to China, and also a deal to grant CNPC a stake in Russian oil fields both onshore and on Russia's Arctic shelf (Marson 2013; Clover 2013). This indicates that China's geostrategic and politicised approach to securing oil and gas supplies continues to dominate the Chinese leadership's thinking on energy security.

In order to achieve these more complex energy goals in a state-led manner that integrates economic, environmental, social and industrial objectives, China's energy institutions required strengthening, streamlining and upgrading. Interestingly France possessed similar institutional and capabilities and policy instruments to those of China in the wake of the 1973 oil shock, notably "a long tradition of state ownership as a tool of national control and exploration", and also chose a neomercantilist policy path entailing state-led energy production (Ikenberry 1986: 111–113). Where such capacity to alter the institutional environment does not exist, for instance in the United States (a classic example of a 'weak' state, following Krasner's (1978) definition, since it asserts a very low degree of autonomy *vis-à-vis* society), whereby policymakers have "few policy instruments of limited range to pursue their objectives" (Katzenstein 1978: 308, 311), policy responses to oil import dependency tend to be more market-driven than those adopted by China and France, despite attempts to strengthen the state's role in the energy sector (Ikenberry 1986: 133). Certainly the long tradition of Chinese state ownership and central planning in the oil sector facilitates China's state-led approach. With reference to the cases of France, the United States, Japan and Germany, Ikenberry (1986: 137) suggests that where the state capacities exist to pursue the statist option, especially during times of strategic adjustment to energy crisis, it is the preferred policy option. The People's Republic of China (PRC) has faced a range of energy challenges throughout its history, starting with the energy crisis that ensued after the Sino-Soviet rift, through to the oil shock of 2007 to 2008. Throughout the Chinese state has exhibited a core statist policy preference, even in the light of more recent attempts at a comprehensive energy policy approach that takes into account the entire energy commodity chain from oil exploration and production through to distribution. The development of this policy preference is motivated by a range of factors that have already been mentioned, but two factors in particular warrant further mention; the socioeconomic features that shape oil policy, and the wider structural context of market transition.

Socioeconomic dimensions of oil policy

As a lifeline or pillar industry, energy clearly intersects with most other policy and economic sectors, and this is especially the case in developing countries such as China. Beijing frequently uses oil policies to advance other social, economic and industrial goals. Since 1993 there has been substantial change within China's oil sector, brought on by the shift to oil import dependency and the massive hikes in world oil prices from 2003 onwards. Policy change during this time was driven directly by the Chinese leadership also in response to wider national economic concerns and requirements, and not just to the priorities specific to the oil sector itself, as Meidan et al. (2009: 606) claim with reference to energy more broadly: "The preference for self-reliance, the importance of economic growth and the consequent rise of energy demand, new strategies for industrial, fiscal and economic reform, and long-standing social policies were all significant factors in the evolution of the energy sector." In a large economy undergoing market transition, national strategic investment in key industries such as energy is considered crucial to economic development and societal stability (Liu 2009: 551). For instance, China's NOCs are tasked with protecting and advancing the national interest in certain ways, even when it sacrifices company profits and efficiency. The most basic socioeconomic responsibility that the NOCs must fulfil is to guarantee a stable and affordable energy supply. In terms of China's attempts to achieve oil supply security the leadership has stressed self-sufficiency and state control of energy production and distribution, both at home and abroad. The preference for state control of the energy commodity chain is not solely a function of concerns pertaining to national security and economic growth, as social equity considerations also play a large role in shaping this orientation. For example, the government maintains control over oil distribution and end-user prices with the aim of protecting energy price sensitive industries, low-income households and disadvantaged communities, and also to help manage inflation, which historically has motivated civil unrest in China. On the policy agenda energy efficiency and environmental protection have increased in significance, especially since 2003 (in response to energy supply shortages, ever-increasing foreign oil dependency, and growing environmental concerns), however, they remain subservient to the immediate demands of economic growth and social stability.

Hence China, as an oil import dependent, developing country faces a set of internal and external challenges that differ from those in advanced western economies, and which have influenced the evolution of the oil sector. Since the Chinese leadership has felt compelled to respond to a wide variety of socioeconomic, industrial, security and political considerations in formulating energy policies, there occasionally appears to be inconsistencies in policy direction, where one policy may even seem to undermine another. For instance, oil price controls that address social equity and industrial concerns appear to run at cross-purposes to directives that target energy efficiency and conservation, as the suppression of oil prices may encourage excessive oil consumption. China's energy policies can also be viewed as misguided or ineffectual when measured against criteria such as economic efficiency. At first glance this could appear to be indicative of incoherent policymaking; however, it is really just a reflection of the numerous priorities that the CCP must juggle in order to ensure economic growth, whilst also trying to address various social, industrial and environmental concerns. These other policy imperatives, particularly the suppression of refined oil prices to help manage inflation, may be viewed as having a negative impact on energy security, but viewed in the contexts of economic development and market transition, where regime survival is heavily dependent on economic growth, the central party-state has somewhat limited options. It is also important to keep in mind that in pursuing the transition to a market economy Beijing decided to adopt a gradualist economic reform agenda, and this has certainly informed oil industry policymaking. During the first Reform Era in particular, gradualist economic reform was shaped by the 'reform without losers' philosophy. According to this view gradualism was a way to implement reform without creating a large number of 'losers', thus maintaining social stability and minimising opposition to the reform process. Importantly gradualism is also a means to build consensus on reforms and enable smooth transitions as new rules and norms are internalised. On the flipside it gives rise to market distortions, which are necessarily addressed through further incremental reform. The wider structural context of gradual market transition in China is thus important to help trace the politics of institutional change in China's oil industry.

Oil industry and firm development under market transition

China's transition from a centrally planned to market-oriented economy provides an overarching structural context within which to trace and evaluate changing state capacities and capacity-building efforts in the oil sector. While the oil sector remains under state control, and will not be fully liberalised any time soon, the party leadership has sought to introduce some market characteristics gradually in order to improve efficiency and performance. It is important to note at the outset that market transition in China has been a heavily state-managed and incremental process. While China has undergone a transition from a planned to a more market-oriented economy over the past thirty years, key features of the Maoist system have been retained, in particular, its political system remains a Leninist party-state, albeit a flexible and adaptive one that has undergone gradual institutional and administrative reform to accommodate the market (Yang 2004). An outside-in appraisal of the evolution of various sectors of the Chinese economy under conditions of market transition might conclude that the market distortions that arise from this gradualist economic transition are indicative of a flawed system. However, the empirical record suggests that such distortions are often temporary and necessary to the transition process by facilitating further reform and restructuring. When viewed from the 'inside out' and in terms of the particular set of policy challenges that China must address, it appears that the Chinese leadership has been quite successful in navigating very complex socioeconomic terrain, and to date has produced policies that have, more often than not, achieved their intended objectives. This is evident in the energy sectors, as Andrews-Speed (2011: 34) notes that the Chinese economy grew 1,500 per cent from 1978 to 2009 accompanied by a 500 per cent increase in energy consumption, hence the government's ability to expand energy at a largely sufficient rate to support such rapidly expanding energy needs is, in itself, quite an achievement. Certainly the gradualist approach to economic transition has produced much better results than the 'shock therapy' or 'big bang' approach to economic reform.

Although China's economy can be considered a hybrid system with many resources being mobilised and allocated through market forces, the form of capitalism that continues to dominate is that where the party-state owns and controls the commanding heights of the economy. Huang (2008: 277) claims the OECD's estimation as of 2003 that the Chinese private sector accounts for 70 per cent of GDP is actually very low by the standard of capitalist economies. Furthermore, China's indigenous private sector relative to the foreign private sector is small, indicating that the selective liberalisation which has occurred favours foreign investors over local entrepreneurs (Harvey 2005: 123). This keeps "the power of the capitalist class offshore" thus inhibiting capitalist class formation within China (Harvey 2005: 123). Since the mid-1990s China's industrial policy has been geared towards consolidating the strategic sectors such as finance and banking, information technology, media, construction, transportation, energy and utilities. Private participation in these sectors is either prohibited or very limited. These dynamics lead Huang to expressly characterise the Chinese economy as a "commanding heights economy" (Huang 2008: 276–277). Beijing's commanding heights approach to economic development allows it to use state industries to serve the country's social and political, as well as economic, objectives. Especially during the second era of reform private enterprises have been placed at a distinct disadvantage in many industries due to the strong financial support the government grants to state firms (Anderlini 2008). Within the private sector the common view over the past decade is that the government is implementing an approach known as 'the state advances as the private sector recedes' (*guo jin min tui*) (Anderlini 2008).

This situation contrasts with the first decade of reform in China where economic development was achieved through relaxing the party-state's control of the economy, decentralising economic decision-making to local government, providing incentives to local entrepreneurs and opening up domestic markets; the predominant trend was indeed towards liberalisation and privatisation (Huang 2011: 4). These trends were reversed through the 1990s, as the party-state reasserted its centralised authority and control of the economy, and concentrated on reform and consolidation of the state sector. Post-GFC China arguably found itself in the midst of the most statist phase of the reform period, with an immense stimulus package of around US$1 trillion poured into the state sector – 90 per cent of which has gone to SOEs (Huang 2011: 4), and the 'massive lending' continued even after the stimulus package ended (Dickson 2011: 42). However, the new Chinese leadership headed by President Xi Jinping and Premier Li Keqiang has pledged to "rebalance the economy and boost the private sector's role in it", which involves weakening state monopolies in the strategic industries (Li 2013; Zhang and Ng 2013). In an attempt to reduce resistance to further state sector reform from large SOEs, corruption investigations of powerful reform-resistant players with connections to the oil industry, dubbed the "petro-purge", were embarked upon by the new leadership (Holland 2013). Whether this is a genuine attempt to push forward state sector reform, or simply an excuse to remove rival political factions and consolidate power, remains to be seen (Holland 2013). Certainly the leadership's current desire to loosen monopolies within the strategic sectors should not be viewed as a move towards liberalisation and privatisation as the party-state "will still retain strong control" (Zhang and Ng 2013).

Hence China's economic development path frames the development of the country's oil industry and NOCs. In other words, the evolution of China's oil industry and the rise of the NOCs have been directly shaped by national economic reform, as well as by reform responding to the specific requirements of this sector. The oil sector in particular is at the forefront of the commanding heights economy, and as such has been subjected to Beijing's continuous drive to reassert central control of the state sector and improve its overall performance. As such efforts to understand China's oil sector development and evolving oil policy approach should be viewed within the context of the country's gradualist market transition. With regard to the oil sector, state capacity has been improved through institutional reforms that were implemented to support the introduction and expansion of market forces, while at the same time maintaining Beijing's control through a range of fairly sophisticated and powerful political, organisational and financial instruments.

Research methods

At a conceptual level the purpose of this study is to contribute to extant literatures and debates concerning the role of the state in economic development, the Chinese policymaking process and the interaction of the main sources of power in China – party/political, bureaucratic and economic – in terms of how they determine the ways in which the state sector is controlled, governed and managed. The proposed method to explore these conceptual themes is a within-case comparison over time, involving analysis of three distinct oil policy regimes in China that have transpired since the PRC's inception in 1949. The historical narrative of China's oil sector development is used to trace the interaction of elite and bureaucratic politics in order to demonstrate how China's oil state capacity has evolved and why particular oil policy approaches were pursued. In comparing these policy regimes attention is afforded to bureaucratic organisation and the role of the CCP in the formulation, implementation and coordination of energy policies, with a specific focus on the oil industry. The changing structural context of market transition and approaches to economic reform are also discussed, since oil sector and NOC development have been influenced by nationwide economic reforms, and not just by factors specific to the industry under investigation. The comparative method is used in order to draw conclusions regarding China's oil state capacity, and shows the influence of market transition in shaping oil policy approaches. In evaluating the contextualised comparative data derived from these cases, insights are deduced via a qualitative methodology. Given China's gradualist approach to economic reform, a historical perspective on the development of China's oil industry is particularly useful. The historical narrative of China's oil sector development reveals the critical junctures or turning points in the political and economic development of the Chinese state. These turning points are of particular importance in Chapter 6, which focuses on how Beijing chose to respond to a range of endogenous and exogenous oil shocks and brings to the fore the importance of state capacity when dealing with critical energy challenges. In this case the country's energy institutions were unprepared and initially lacked the ability to respond effectively, leading Kong (2006) and others to speak of institutional insecurity as the major risk to energy security in China. Ikenberry (1986: 106) sums up the influence of extant state capacities on policy outcomes by stating; "at moments of crisis and change, as during oil shocks, the distinctive structure of the state itself shapes and constrains the substance of strategic policy". Such moments of crisis also provide the impetus for strategic adjustment, where political elites may instigate institutional reform and change. In terms of China's oil sector development, critical junctures, such as an energy crisis, provided the impetus for the party leadership to restructure the state sector. However, institutional change has not only occurred during times of crisis, but also during 'normal institutional life' where we see incremental periods of change in a wider context of social, economic and political relations (Bell and Feng 2013: 32–38).

Chapters 4 to 7 situate the conceptual analysis undertaken in Chapters 2 and 3 in an empirical context. The distinct value of employing qualitative methodology in comparative politics research has been discussed extensively since the 1990s. In contrast to trends in the 1970s, which saw qualitative research as a last resort technique to be used when quantitative methods fail or were considered inappropriate, many scholars now believe qualitative methods are essential for addressing particular research problems and existing conditions of knowledge (Mahoney 2007: 122). Ragin (2000: 44) notes that qualitative methods often compel researchers to reconceptualise cases, and reconsider causes and outcomes. The prevailing approach followed in this project does not, as with most quantitative approaches, consider 'single-shot' hypothesis testing the central task in knowledge accumulation. Such determinate research designs often preclude theoretical innovation. Rather, as Collier et al. (2004) argue, research design should involve an interactive process between theory and evidence. Collier et al. (2004: 238) advocate the production of interpretable findings resulting from 'a particularly revealing comparative design' and 'a rich knowledge of cases and context', which can plausibly be defended. In an interpretive context, the qualitative assessment of case studies and contextualised comparisons are not second best choices; rather they constitute appropriate responses to the existing conditions of knowledge about the subject to be studied, in this case political elites, government bureaucracies and SOEs.

Organisation and chapter synopses

In sum, the main objective of this study is to address the overarching question of how state capacity and governance dynamics within

China's oil sector impacts policy formulation and implementation, ultimately shaping the country's domestic and international energy behaviour. In doing so, it also addresses contrasting debates about China's political future, specifically whether the country's authoritarian regime can effectively adapt and respond to various governance challenges. This volume essentially argues that China's state capacities are successfully adapting and strengthening, and that the country's ability to effectively formulate and implement energy policies in the oil sector, and manage the NOCs, has improved since the early 1990s. In other words, control over the oil industry by China's political elites is greater than what tends to be implied by the FA model. The role of the CCP is central to understanding policymaking in China, yet often fails to be adequately accounted for in the extant models that explain the Chinese policy process – the FA model in particular tends to ignore or downplay the party's tremendous political power. This power is in itself a major counter to fragmentation in the political system. Hence in addition to demonstrating that governing capacity in China's oil sector has improved over time, the role of the CCP is shown to be central in understanding how these changes in state capacities have come about. This study has adopted a straightforward structure, whereby the conceptual framework is first explained, and subsequently followed by four case study chapters, the first three of which unfold chronologically, with the final case chapter shifting the focus from the institutions used to govern the oil sector to the relationship between the Chinese government and the NOCs.

The theoretical framework hinges on the concept of state capacity and also the interplay of elite and bureaucratic politics with reference to models of the Chinese policy process. These are discussed in Chapters 2 and 3. In Chapter 2 the state capacity concept is unpacked and its relative strengths and weaknesses are assessed. This is followed by a more thorough evaluation of the extant literature on China's state capacity, which is essentially split between two different camps, termed the 'optimists' and 'pessimists' by Shambaugh (2008: 24–25). Chapter 3 examines the relationship between elite and bureaucratic politics in China, which is central to understanding and evaluating state capacity and how it has changed over time. The FA model is explained and critiqued, and the importance of the CCP is explored and emphasised as the key variable that explains institutional change in the oil industry. Hence this chapter argues that the focus of analysis when evaluating state capacity in the oil sector needs to incorporate political elites, and not just government bureaucracy and administration, advancing BA as a more appropriate model to understand the policymaking process in China.

Chapters 4 through 6 examine institutional developments and policy approaches in China's oil sector since 1949. These are broken up into three eras: (1) The pre-reform autarkic phase of oil industry development which lasted from 1949–1977; (2) A period running from 1978 to 2002 of increasing fragmentation and decentralisation, which also saw processes of marketisation and corporatisation transform the oil industry; and (3) A more focused and coordinated phase that began in 2003 witnessed institutional arrangements gradually become more sophisticated and integrated, and the central party-state strengthen and expand its political, organisational and fiscal capacities. In each era the political and institutional landscapes are closely examined in order to explain both policy articulation and policy outcomes. The last two eras are of greatest interest to this study, since a major focus is to explain both how and why China has responded to a range of endogenous and exogenous energy challenges from the beginning of reform to the present day. However, institutional development in earlier decades is also explored as it helps us to understand contemporary governance dynamics through identification of the historically accumulated arrangements of Chinese bureaucratic and party institutions.

The focus of Chapter 6 is on those institutional developments since 2003 that have improved state capacity not only within the oil sector, but also within the Chinese state more broadly. China's shift to the status of net oil importer in 1993 saw the development a statist oil policy approach based on a preoccupation with the security of oil supply under conditions of bureaucratic fragmentation and decentralisation in the oil industry, with the NOCs filling the policy vacuum. However, from 2003 onwards more coordinated and effective energy policies have emerged as the top leadership turned its attention towards energy security and implemented a range of capacity-building reforms (both political and bureaucratic). In this chapter central energy policy directives and institutional objectives are evaluated against energy policy outcomes within the oil sector. Specific examples include the success of SASAC, and other relevant government agencies, in producing internationally competitive NOCs, and also how the central party-state has been able to balance oil sector priorities with other socioeconomic and industrial considerations, which became a particularly urgent task from 2007–2008 when international oil prices skyrocketed. Arguably more important than the oil sector's bureaucratic structure, which continues to present some challenges to policy-making, are those key policy, fiscal and organisational instruments that remain under party-state control, notably oil pricing, the NOCs, and the banking system and investment approval system (China's financial institutions are required to provide credit and foreign exchange support to the NOCs to undertake investment activity in strategic locations). These tools of influence and control are effectively deployed to ensure compliance with central party-state policy directives.

The international dimension is also noted throughout and draws upon the inside-out mode of the analysis to explore how state capacity in China's oil sector has influenced the country's international energy behaviour. The activities of the NOCs are the primary focus in this regard, taking into account the nature of their operations and the support that is provided by the Chinese government to promote domestic and international energy investments. Most importantly, the nature of the relationship between the Chinese government and the NOCs is analysed in Chapter 7. In this chapter the corporate governance structures that the NOCs operate under are evaluated, showing how they perpetuate government control of state firms. Finally, contrary to mainstream commentary, China's international oil activities are shown to be a function of a fairly coherent energy policy approach and outward direct investment strategy called the 'go global' policy, which is implemented by various state bureaucracies including the NOCs, the China Eximbank and CDB, and the Ministry of Foreign Affairs (MFA). This policy, which also addresses Beijing's much discussed oil diplomacy, is a focus for the latter part of Chapter 7. Here it is concluded that China's attempt to integrate its international energy policy with economic and foreign policy has been reasonably effective. The NOCs enjoy direct and indirect economic and political support from a variety of government channels, which has gone a long way toward helping them become internationally competitive in a relatively short period of time.

The conclusion (Chapter 8) locates this study within ongoing debates about China's political future. The evidence gleaned from the empirical chapters indicates that the central party-state is reasonably flexible and adaptable, and has strengthened its ruling capacity over the past two decades. Rather than stagnating or crumbling under the forces of globalisation and liberalisation, like most other communist single party-states, China's authoritarian regime is indeed resilient, and able to respond reasonably effectively to a wide array of social, political and economic challenges. The CCP has developed the tools and tactics that grant it the ability to learn, adapt and reform, hence many of its policies tend to be proactive and offensive, rather than reactive and defensive. These trends counter the prognostications of the modernisation and democratisation literatures, as these suggest that democratic transition is more or less inevitable, once a certain level of political and socioeconomic development and organisation has been achieved.

2
Sectoral Governance and State Capacity

The concept of state capacity is central to understanding the ways in which China's quest for energy security has manifested itself, both domestically and internationally. Specifically, the nature of the political authority and institutional arrangements that govern China's oil sector need to be evaluated in relation to how they enable a state-led approach to oil production. Such an exercise enhances our knowledge of how China governs the strategic sectors of its economy more broadly, as these sectors remain under party-state control. In examining oil state capacity this study addresses two contrasting characterisations of China, commonly referred to as the 'fragmentation thesis' and the 'rise of China' or 'China threat theory', which have both proved to be remarkably influential and enduring in scholarly and policy circles (Naughton and Yang 2004: 2–5). The former defines China's strength in relation to state capacity and the exercise of political authority. Broadly, it concludes that the central party-state's steering capacity, that is, the ability to guide national socioeconomic development, is greatly hindered by both economic liberalisation, and political, administrative and fiscal decentralisation. Since the beginning of the Reform Era these processes have allegedly resulted in a retreat of the party-state's role in economic governance, the proliferation of corruption among political and business elites, and a fragmentation of authority among multiple government units leading to bureaucratic infighting, policy inertia and coordination failures. Furthermore, the CCP's capacity for proactive adaptation to changing circumstances and its ability to effectively respond to socioeconomic problems is also considered fairly low. China experts who adhere to this position essentially see the Chinese political system as severely "embattled and endangered" (Shambaugh 2008: 25).

On the other hand, those concerned with China's rise in world affairs, focus on the country's 'external posture', including its rapid and sustained economic growth, military modernisation and relatively skilful diplomacy. Taking a view from the 'outside-in' they typically argue that the Chinese state is actually very strong, and in possession of a centralised and coherent decision-making authority (Shambaugh 2008: 24). Arguably each position is far too one-sided and overstated, and often sensationalist in tone (Naughton and Yang 2004: 2–5; Shambaugh 2008: 23–25). However, despite their significant shortcomings, these contrasting perspectives also prevail in the literature on China's oil policymaking and energy security. With regard to China's domestic oil sector, experts such as Downs (2004a), Lester and Steinfeld (2007), Zha (2006), Kong (2006 and 2010) and Meidan et al. (2009) have focused on the supposedly crippling effects of fragmented energy policymaking and implementation authority and, among other things, the inefficiencies of state-led oil allocation and pricing. Conversely, when attention turns toward international energy behaviour, the strength and coherence of the Chinese state is drastically over-emphasised, with its neomercantilist approach to acquiring oil resources abroad often touted as a direct threat to international oil markets and other oil importing countries. In these accounts scant attention is afforded to the oil challenges that exist within the Chinese state, and how oil policy is also used to further other social, industrial and economic aims and objectives. Elements of these contrasting views of China's state capacity are useful to understand the strengths and weaknesses of the Chinese state, but alone they provide an incomplete and inaccurate explanation. This study attempts to provide a balanced and nuanced explanation of state capacity in China's oil sector that occupies a middle ground between these positions.

While state capacity constitutes the core concept, this study is further nested theoretically with reference to models of policymaking in China that shed light on flows of political authority (top-down and bottom-up) within the Chinese political system, namely FA and BA. A detailed account of these policymaking models is provided in Chapter 3. Of particular interest in the context of this study is state capacity *vis-à-vis* China's strategy in governing the state sector of its economy. Strategic intent and coherence, as well as the authority of the central government to execute policies, are key to assessing China's oil state capacity. Hence the analytical focus is on the capacity of the state to strategically and effectively steer the economy, rather than on other traditional criteria for examining state capacity, such as fiscal extractive capacity. In addition, the legitimacy, coercive and adaptive capacities of the centre, which lend themselves to effective policy formulation and implementation, are examined. In the case of China's oil sector this volume challenges the dominant view among political scientists that there is a lack of coordination and coherence in terms of energy policy formulation, combined with a chronic inability to effectively implement energy policies in a consistent manner. Rather, it is argued that the effectiveness of China's energy policies has increased over the past decade, and that the political authority and capacity of the central government to implement its policy decisions is relatively strong. Indeed in the final analysis it is clear that, in most instances, energy policy objectives have been met. Since the early 1990s, the central party-state has been well aware of the need to strengthen its governing capacity and has taken many significant and effective steps toward realising this goal. This is not to say that the extant conditions of bureaucratic fragmentation and decentralisation do not pose a challenge and have some influence on the way policy is formulated and implemented, but rather that the authority of the centre to impose its will on the state's policy and administrative apparatus greatly improved, and therefore should not be underestimated. Furthermore, the trend of incremental political institutionalisation in China over the past two decades, whereby the CCP has evolved away from the personalised power of the party leader (as was the case with Mao and Deng) towards an increasingly consensus-building collective leadership, has conferred greater stability and predictability to the policy process in general, enabling the state to penetrate society and implement policy much more evenly and effectively (Miller 2008a: 61–62).

This chapter is divided into three parts. First, the state capacity concept is defined and unpacked, and the analytical advantages and disadvantages of using this concept are explored. Second, the relationship between state capacity and state autonomy (the independence of the state itself, which determines how impervious it is to external (social) causal influences), of both despotic and infrastructural forms, is examined, since this is necessary in order to understand the nature of political authority and capacity possessed by the central party-state. Lastly, extant views of state capacity in China are presented and evaluated, with particular attention afforded to seminal and contrasting works by Pei (2006a) and Shambaugh (2008), along with works by many other China experts. What becomes clear when looking at the striking diversity of views on the state of China's state capacity is that evidence can be found to support almost any generalisation, due to the sheer size, pace of change and heterogeneity of the Chinese state. This supports the case for disaggregating 'China' by pursuing sector specific research, rather than attempt to generalise the Chinese political system and economy writ large.

The concept of state capacity

State capacity simply refers to the ability of the state to formulate and implement policy, in other words, "...to transform its own will and purposes into reality" (Wang and Hu 2001: 5). It is widely perceived by analysts as key to explaining diverse processes and outcomes such as differential patterns of economic growth and development, democratisation, the provision of public goods, variations in state surveillance, nationalism, intra-state conflict and international wars (Soifer and vom Hau 2008: 220). Researchers typically focus on key institutional foundations of state capacity such as the degree of state autonomy, and the existence of a bureaucracy that enables purposeful and coordinated action, adequate resources and appropriate policy instruments (Bell 2002: 10). High capacity regimes are generally considered better equipped to establish a monopoly on the use of force, enforce contracts, maintain social order and stability, achieve positive development outcomes and higher rates of growth, regulate institutions, combat corruption and extract and allocate financial resources (Soifer and vom Hau 2008: 220). Hence state capacity is recognised as a necessary condition for the strength and economic well-being of the nation-state. In this study the capacity of party elites to restructure and reorganise institutional arrangements within the Chinese state is considered central to explaining institutional development and change in China's oil sector. The key dynamic has been the strengthening and expanding capacity of the central party-state to exert top-down authority during the second era of reform, which has enabled the centre to push through its policy agendas and override bureaucratic conflicts or inertia.

It is useful to identify the types of capacities that enhance the overall effectiveness of the state in formulating and implementing its policies. Wang and Hu (2001: 25) describe four interrelated capacities that determine the ability of the Chinese party-state to realise its policy goals: fiscal extractive capacity, steering capacity, legitimacy capacity and coercive capacity. Every state will possess these four capacities to varying degrees. In addition, adaptive or learning capacity is occasionally cited as another important element of state capacity. Traditional notions of state capacity tend to revolve around the state's ability to levy taxes to finance public goods (Levi 1989). Historical sociologists such as Brewer (1989), Hoffman and Rosenthal (1997), and Tilly (1985) argue that the imperative of taxation developed in response to the exigencies of war, as those governments that built fiscal capacity were more likely to win wars and less likely to face opposition from their citizens. North (1981: 21) went so far as to define the state itself in terms of taxation powers when he said that the state is "an organisation with a comparative advantage in violence, extending over a geographic area whose boundaries are determined by its power to tax constituents". Hence fiscal extractive capacity is typically considered the most fundamental and important measure of state capacity as it provides a basis for the realisation of other state capacities (Wang and Hu 2001: 5).

The state's steering or regulatory capacity refers to the ability of the state to guide national socioeconomic development. It entails the mobilisation of financial resources to "carry out resource allocation...[and] stabilise the economy" (Wang and Hu 2001: 27). Wang and Hu (2001: 27) refer to legitimation capacity as "the state's capacity to regularise the behaviour of individuals in society, and to utilise the financial resources at its disposal to regulate any contradictions among interest groups in society, to allocate economic interests fairly, and thus to realise the goals of social development." Coercive capacity refers to the central government's power to enforce policy compliance and obedience. Legitimation capacity entails "the capacity to dominate by using symbols and creating consensus" (Wang 1995: 89). The legitimacy of the CCP is now largely performance-based, and is perceived by some to be threatened by widening inequality and rampant corruption (Wang 1995: 108). Adaptive or innovative capacity is another significant element of state capacity, and refers to the ability of a political system to proactively adjust and adapt to new circumstances. Heilman and Perry (2011) propose the concept of 'adaptive governance' to describe the capacity of the Chinese state to respond effectively to a variety of governance challenges. They conceive of China's policymaking as "a process of ceaseless change, tension management, continual experimentation, and ad hoc adjustment" (Heilman and Perry 2011: 3). This dimension of state capacity is particularly significant for single-party regimes, which are arguably prone to stagnation and ossification due to conditions of more limited political pluralism and competition. Tsai (2007: 8–9) notes that within the authoritarian context, institutional change usually only occurs through implosion or revolution, or in response to external shocks rather than endogenous pressures. By contrast, China has shown itself to be remarkably adaptive, flexible and proactive in instituting reforms both within the Chinese state and within the party itself during both crisis and non-crisis periods.

In terms of its utility as an analytical tool the concept of state capacity has been criticised for being difficult to measure and too broad to function as an adequate framework for country case studies. It is indeed a multi-dimensional concept that needs to be disaggregated in order to be operationalised. Some aspects of state capacity are easier to quantify and measure than others. For example, measuring fiscal extractive capacity is fairly straightforward. However, most other types of state capacity are not so tangible, and tend to be measured retrospectively through observation of their consequences, mainly in terms of policy outcomes (Kjær and Hansen 2002: 7). In addition to disaggregating to notion of 'capacity', it is also necessary to disaggregate the 'state' when thinking about 'state capacity'. The concept of state capacity frequently implies that the state is a uniform and homogenous entity. With reference to this assumption, Weiss (1998) poses the question "Capacity for what?" to further her argument that the modern state is not a monolithic or coherent unit, hence analysing it at an aggregate level will yield few useful insights. Rather, it is more productive to acknowledge that state capacity can vary dramatically across state functions and policy areas, and should be analysed accordingly.

Ultimately the concept of state capacity is based on the premise that 'states matter' when analysing socioeconomic development. This idea of the state as a distinctive actor that shapes societal outcomes has of course been challenged in more recent times by trends such as economic globalisation and the spread of neoliberalism, postmodernism, the new regionalism of the European Union, the third wave of democratisation, and the purported emergence of a global civil society. In relation to these phenomena Weiss (1998: 2) identified a "state denial" zeitgeist that emerged across the social sciences, which asserted the imminent demise of the state's importance in "structuring social relations", and its looming obsolescence as an 'organising principle'. The notion of growing state incapacity as a result of neoliberal policy convergence gained a strong foothold in the fields of comparative and international political economy during the nineties (Weiss 1998). According to this view national governments were perceived as increasingly powerless to control and regulate their domestic economies and implement effective social policies. In *The Myth of the Powerless State*

(1998: 3) Weiss argues that this view is inadequate in that it remains "quite blind to the variety of state responses to international pressures, and to the sources and consequences of that variety for economic prosperity." Weiss (1998) argues that the capacity of the state to adapt and develop strong domestic institutions is actually the critical determinant of industrial competitiveness and domestic adjustment to the global economy. Hence state capacity matters more rather than less in the face of the strong external forces of economic globalisation. Academic attention has certainly shifted back toward the role of the state and bureaucracy, particularly within the field of development studies, whereby state capacity building is considered a prerequisite for initiating and sustaining economic growth and development. In policy terms the focus has also shifted back to the role of the state in economic development. For instance, capacity building is now advocated by the Post-Washington Consensus as fundamental to achieve positive economic growth and development outcomes, given the failure of the neoliberal Washington Consensus to deliver sustained growth in a range of developing countries. This is especially the case given the perceived failure of neoliberal prescriptions of "rolling back the state" in favour of market forces during the 1980s and 1990s.

State capacity and autonomy

While state capacity refers to the ability of the state to realise its goals, state autonomy dictates the state's freedom to choose them and is "necessary for effective state intervention" (Wang 1995: 90). Hence the relationship between state capacity and autonomy captures the nature of the state-society relationship and its effects on policy formulation and implementation. This is neatly summarised by Bell (2002: 10): "State autonomy implies a degree of insulation from societal pressures and political opponents which in turn implies that such states have the capacity to push through reforms, despite opposition." States that possess a high degree of autonomy are able to push through "pluralist political gridlocks" and get things done (Bell 2002: 10). With regard to China the strength and autonomy of the state *vis-à-vis* society is considered a significant analytic element. China is an example of a 'strong' state, which Krasner (1978: 55–56) defines as "easily able to not only resist societal pressures, but also fundamentally alter the societal environment". In conceptualising the sources of state autonomy from civil society and the differences between 'strong' and 'weak' states, Mann

(1984) identified two dimensions of state power: despotic and infrastructural. Despotic power refers to power 'over' society, where rulers can attempt to implement policy "without routine, institutionalised negotiation with civil society groups" (Mann 1984: 188). Infrastructural power, on the other hand, is power 'through' society, referring to the state's capacity to reach down into society and implement policies successfully "throughout the realm" (Mann 1984: 188). According to Kjær and Hansen (2002: 12) Mann's distinction also illustrates how low capacity states will often rely upon despotic rather than infrastructural powers, due to "their inability to build effective state administrations and to impose centralised administration over a fixed territory." Modern authoritarian regimes often enjoy both high despotic power and infrastructural power (China being a good example), as policies can be formulated and promulgated in a despotic manner and implemented successfully due to the presence of sufficient infrastructural power. Mann (1984) further posits that this combination of high despotic and infrastructural power in modern authoritarian states confers greater stability, which may prolong the survival of these regimes.

When looking at China's oil sector it is important to recognise that all the key actors involved in China's oil policymaking, including the NOCs, are located within the party-state sphere. Hence the absence of direct influence or pressure from societal actors is even more marked within this sector. Although there are no domestic societal actors that directly influence China's behaviour with regard to the oil sector, this does not mean that societal pressures and expectations have no effect on policymaking. In this sense it is very difficult to separate state and society realms completely. Studying either the state or society as completely autonomous units is not conducive to capturing the range of indirect constraints that both entities place upon each other, even in a single-party state characterised by authoritarian rule. The contemporary social contract that governs the state-society relationship in China is no longer convincingly based upon ideological communism, but rather on economic performance, the responsiveness of the party-state to internal and external challenges, and appeals to nationalism. These foundations for domestic legitimacy have an impact in terms of shaping China's policy orientations, and provide an overarching context whereby China's lack of adequate domestic oil supply, which could severely compromise future economic growth and development, poses a clear threat to the survival of the party-state. This elevates energy policy to the realm of high politics, which is evident in former President Hu

Jintao and Premier Wen Jiabao's active involvement in securing foreign oil supply, a trend continued by the new president Xi Jinping, and further motivates the preference for a statist approach to energy security. Therefore, it is worth noting that the domestic social environment can broadly influence the preferences of policymakers.

Traditional statist conceptualisations of state capacity adhere to a zero-sum view of the state-society relationship, whereby capacity is considered to depend upon the state's ability to overcome political and societal constraints, such as factional and interest group opposition, when pursuing its goals (see, for instance, Migdal 1988). The sources of such autonomy can often be found in the weakness of non-state actors, and also in specific social norms favouring collective action in the national interest, which are present in some political cultures, notably in the East Asian region (Polidano 2001: 21–22). During the 1980s the first generation of developmental state theorists focused on the role of state autonomy in East Asian development (see, for instance, Johnson 1982). Specifically, they credited rapid economic growth and development in Japan, South Korea and Taiwan to strong autonomous governments that were able to set developmental goals and discipline weak industrial sectors via direct intervention. These statist explanations of the 'East Asian miracle' coincided with the rise of the "bringing the state back in" school, spearheaded by Evans et al. (1985), which had gained currency within political science. By the late 1990s a second generation of development theorists challenged this developmental statism by advocating a relational perspective. They rejected the notion that high capacity developmental states are autonomous in the sense that they are completely insulated from societal actors, instead arguing that their developmental success is actually derived from connectedness to society. This approach is captured in Evans' concept of 'embedded autonomy' (Evans 1995). This relational capacity conceived in terms of state-society 'embeddedness' is not relevant to the case of China's oil sector as all the key actors that are involved in oil policy articulation and implementation are located within the party-state sphere. Hence rather than focus on relational capacity, this study analyses institutional capacity within the Chinese party-state.

Contrasting views of state capacity in China

The Tiananmen Square protests and collapse of communism in the Soviet Union and the Eastern bloc in the late 1980s spawned a vast literature on China's political future and the likely prospects for the

CCP's survival as the ruling party. The overwhelming consensus in the immediate aftermath of these events was that China's regime would soon follow suit and fall to the third wave of political liberalisation and democratisation (Nathan 2003: 6). Twenty-five years later scholars continue to debate the political trajectory of the Chinese state, but now grapple with the reality that, far from teetering on the verge of political collapse and transformation, China has strengthened the resilience of its brand of authoritarianism. Much of the scholarly work on the current condition of, and future prospects for, the Chinese party-state relies upon critical evaluations of state capacity. Within this academic terrain there is a reasonable degree of consensus over the various governance problems and systemic maladies that afflict China (Shambaugh 2008: 24). However, beyond this there is stark disagreement about the severity of these governance challenges, and the effectiveness of the party-state's responses to them (Shambaugh 2008: 24–25).

The recent scholarship addressing this subject matter is polarised between those scholars who are highly critical and sceptical of the ability of the party-state to deliver long-term economic development and build effective institutions, and those who focus on the party-state's flexibility, adaptability and resilience, referring to a range of successful state-led reforms. The more pessimistic appraisals arguably underestimate China's ability to adapt and respond to a daunting set of governance challenges. Bell and Feng (2013: 19) claim that the "contradictions associated with China's experience of liberal reforms under political autocracy" have polarised the literature into these two camps. These contradictions include "double-digit economic growth based on an unsustainable development model; internationally renowned achievements in poverty reduction combined with growing income inequality; state-led institutional reforms on a sweeping scale plagued by rampant micro-level corruption; and a dynamic private sector combined with continuous dominance by state enterprises in key sectors" (Bell and Feng 2013: 9). Kroeber (2008: 29) attributes the vast collection of differing viewpoints pertaining to almost every aspect of Chinese political economy, to the country's sheer scale, complexity and rate of change, which means that it is possible to find evidence that supports almost any generalisation. This is especially the case in governance studies undertaken at the aggregate level of the Chinese state. Hence in order to achieve a degree of theoretical parsimony and lucidity, selectivity is a necessary requirement. Disaggregating 'China' by pursuing sectoral research, instead of treating it as a monolithic entity, can enhance explanatory power and accuracy.

Of those assessments of China's state capacity that have been produced over the past two decades (since the shock of Tiananmen), a significant proportion have been bleak in their outlook. There has been a clear tendency among many political scientists, notably Chang (2002), Shirk (2007), Walder (2004), Gilley (2004), Baum (1991), Walter and Howie (2011), Pei (2006b) and Li (2012), to provide explanations and prognostications that focus on the corrupt, weak, inefficient, fragmented and despotic aspects of China's political system. The conventional view held by most of these scholars is that there has not been any significant political reform to match the momentous economic reforms that were unleashed during the Reform Era, and this has resulted in institutional decay and a dramatic rise in systemic corruption. Naughton and Yang (2004: 2) claim that these sorts of analyses are one-sided and informed primarily by the influential and enduring 'disintegration' or 'fragmentation' thesis. This thesis initially emerged in reaction to the economic and political decentralisation trends of the 1980s and the crisis at Tiananmen, and has continued to gain currency over the past two decades (Naughton and Yang 2004: 2). Naughton and Yang (2004: 2) provide a neat summary of the main thrust of this body of literature; "Focusing on the fissiparous forces that had developed in China following more than a decade of decentralising reforms, this literature extrapolated from the trends of the 1980s in order to look into the future, and it predicted an increasingly decentralised, unregulated, and ultimately uncontrollable society. At the same time, under the influence of the harsh crackdown that occurred at Tiananmen, these authors tend to be profoundly pessimistic about the ability of the Chinese government to adapt to and cope with these trends."

Within the extant literature there is little dispute that the decentralising and liberalising economic reforms of the 1980s resulted in a steady decline of the central party-state's control over the economy, society and aspects of hierarchical political authority, where local governments were empowered at the expense of the centre. This retreat of the central government was also accompanied by the proliferation of corruption and a decline in state effectiveness in both the political and administrative realms, particularly since China "abandoned crude but powerful tools of government resource allocation before market-friendly indirect and regulatory institutions were available" (Naughton and Yang 2004: 1). In the light of these alleged signs of disintegration and state incapacitation, some scholars, notably Chang (2002), have gone so far as to predict an imminent and inevitable collapse of the Chinese state followed by democratic transformation. Others such as Gilley (2004) and Zheng (2004) pursue the notion that a 'democratic breakthrough' would soon occur in China. Arguably these scholars have misunderstood the causes and overall strategy behind this seeming decline of party-state control. While the CCP has clearly been in a "progressive state of atrophy for many years...in terms of its control over various aspects of the intellectual, social, economic, and political life of the nation" (Shambaugh 2008: 3), this has been part of a deliberate strategy of loosening the party-state's grip on many part of economy and society, while strengthening it in several key, strategic areas.

Shambaugh (2008: 26) makes the observation that a large number of Chinese émigré scholars perceive an "increasingly incapacitated party-state" beset by political decay. Among this group Pei offers a particularly harsh and notable analysis of China's political development. In *China's Trapped Transition: The Limits of Developmental Autocracy* (2006b) Pei argues that the combination of gradualist market reforms and single-party rule creates contradictions that are increasingly unsustainable, and are beginning to have deleterious effects on the significant socioeconomic gains that China has made over the past thirty years. During the course of his study, Pei exposes the weaknesses of the reform strategies undertaken since the 1980s. These reforms have given rise to a variety of problems, the most significant of which include the increase of endemic and systemic official corruption and rent-seeking behaviour, the decline of state capacity and the growing structural imbalances in Chinese society and polity. Pei (2006b) claims that a 'partial reform equilibrium trap' has emerged whereby the decentralised and self-serving or 'predatory' party elites focus on their own interests and maximise their own gains rather than pursuing developmental goals, hence at the expense of Chinese society and economy. According to Pei this trapped transition is likely to manifest as a period of prolonged political and economic stagnation.

A significant aspect of *China's Trapped Transition* is Pei's damning critique of gradualism. While gradualist reform may have served China's development well in many respects, Pei (2006b) argues that the political logic that underpins it sets up a 'self-destructive dynamic' within the Chinese state. Under conditions of slow gradualist reform within authoritarian regimes, free market forces cannot determine economic outcomes, and as a result there are imbalances and distortions that arise which can be readily exploited by those in power. In other words, where there is low political accountability and transparency slow economic reform gives rise to opportunities for rent-seeking behaviour and other corrupt practices. Although Pei (2006b) attempts to show that gradual reforms have produced this self-destructive dynamic in the interim, there is little evidence to suggest that reform momentum will necessarily cease or stall and trap China in this stage of development. Pei's arguments regarding political and economic stagnation and state incapacitation do not bear scrutiny given China's impressive record of both economic growth and strengthening state capacity to deliver public goods. More importantly for the purpose of this study, Naughton (2008a: 126) argues that Pei's analysis fails to recognise that the "decisiveness of the Chinese government has increased during the reform process... There is no evidence of paralysis in decision making...since 1993, the government has consistently enacted important reforms once the top leadership has recognised a compelling national interest." The major flaw in Pei's thesis is not so much his analysis of the systemic maladies that afflict China at present, notably the rise of crony capitalism and rent-seeking behaviour, but a vast underestimation of the ability of Chinese political elites to respond effectively and keep the reform momentum going. This flaw is endemic to many of these more pessimistic predictions of China's political future. Similarly Li (2012) admonishes China's resistance to democratisation, and argues that the Chinese state is actually a 'stagnant system'.

On the other side of the debate there is a growing group of scholars who offer more balanced, and in some cases optimistic, appraisals of China's state capacity. A starting point for many of these studies is the observation that China has managed not only to survive, but also to adapt to new circumstances and grow stronger, especially during the second Reform Era, whereas most other communist party-states throughout the world have imploded. Leading scholars on this side of the fence include Yang (2003 and 2004), Nathan (2003), Miller (2008a), Naughton (2008a), Dickson (2003 and 2011), Tsai (2007), Heilman and Perry (2011), and Bell and Feng (2009 and 2013). For example, Nathan (2003) and Miller (2008a) focus on institution building that has occurred within the CCP during the Reform Era. They argue that several key elements of the institutionalisation of Chinese elite politics have been established with considerable success, the most important being the normalisation of leadership succession, which is fundamental to achieve long-term regime stability. Miller (2008a: 61) further examines the incremental institutionalisation of elite politics in China, which has seen a move away from the leader's personalised power, as embodied by Mao and Deng, towards consensus-building collective leadership. She argues that this has produced a more stable, routinised and predictable policymaking process, which is "indispensable to the success of China's modernisation" (Miller 2008a: 62). The institutionalisation of political succession was exemplified by the relatively peaceful, rule-bound and orderly power transitions from the third through to fifth generation of Chinese leaders. Nathan's (2003: 9–10) 'resilient authoritarianism' thesis also identifies the increase of meritocratic promotion of political elites, the functional specialisation of institutions and policy portfolios staffed by competent technocrats and the establishment of institutions for political participation (input institutions), as significant factors that enhance the internal stability and legitimacy of China's authoritarian regime. All of these developments are deemed to have increased the regime's ability to detect socioeconomic problems, and enables more effective policy responses to public needs. By emphasising the CCP's social responsiveness and ability to provide public goods, Nathan challenges the notion that the party is suffering from a chronic loss of political legitimacy and is unable to govern effectively. Likewise, Heilman and Perry (2011) emphasise the resilience, adaptability, flexibility and innovative capacity of the Chinese state, proposing the concept of 'adaptive governance'. Interestingly they trace the practice of adaptive governance to 'guerrilla-style policymaking', a Maoist experimental method that emerged during the communist revolution that unfolded in China in the 1930s and 1940s. This style of policymaking was a fluid and "continual process of improvisation and adjustment," where local initiative and experimentation are encouraged, but strategic decisions are the domain of the top Chinese leadership. Hence China's leaders 'never let slip the reins of power' and have remained in the driver's seat when it comes to strategic decision-making and determining the overall direction of the economy, but they also encourage much local initiative, which can be readily undone or annulled if it does wrong (Heilman and Perry 2011; Mirsky 2012).

Moving beyond a concern with the strength of the party, Yang (2003 and 2004) seeks to examine the strength of the broader Chinese state, and as such deals comprehensively with China's growing state capacity, particularly in the area of economic governance. In his seminal work on China's reform of the institutional framework for economic governance, *Remaking the Chinese Leviathan: Market Transition and the*

Politics of Governance in China (2004) Yang argues that rather than weakening, as is commonly assumed, central party-state power and authority in China is actually strengthening, which is the result of deliberate efforts by China's leaders to remake the country's institutions of governance and, in particular, reconstitute the "sinews of central governance." Importantly, the Chinese leadership has rebuilt its fiscal sinews, which had been weakened by "particularistic arrangements" involving fiscal contracting between the centre and the provinces during the first decade of reforms (Yang 2003: 44). Comprehensive tax and fiscal reforms were launched in 1994, creating a modern and recentralised tax-assignment system. Furthermore, the central leadership has also attempted to remake the state to better suit market conditions and maintain market order, especially by tackling corruption and setting minimum standards. To this end, the first comprehensive and coherent reform drive lasted from 1998 to 2003. These reforms bore results in two main areas: the downsizing, rationalisation and specialisation of the government, and the establishment of regulatory agencies. According to Yang, these reforms have laid the foundation for a modern regulatory state in China. The major Soviet-style line ministries that constituted the core of the planned economy were abolished, and replaced with regulatory institutions that have adopted "vertical administrative methods to promote agency integrity" (Yang 2003: 46). Despite being a work in progress, Yang claims that to date these initiatives to improve the efficiency, transparency, and accountability of the administrative state have been fairly effective, though they have been unevenly adopted across functional and geographic areas. China's successful adaptation to accommodate the emerging market economy, has led both Ramo (2004) and Halper (2010) to go so far as to argue that China's state-led capitalist model now serve as a credible alternative to the Washington Consensus.

A particularly important study that acknowledges the centrality of political elites to institutional change and the policymaking process in China is Shambaugh's *China's Communist Party: Atrophy and Adaptation* (2008). Shambaugh persuasively argues that although the CCP, as an institution, has been in a state of atrophy for many years in terms of its control over society, its authoritarian resilience and endurance has increased as a function of an adaptable, flexible and increasingly responsive party-state elite. Shambaugh demonstrates that the CCP is readily adapting to meet a comprehensive set of challenges that arise both from single-party rule, such as leadership succession, and also from the pressures associated with socioeconomic modernisation and globalisation, such as increasing social stratification, the need to deliver public goods, rising corruption, demand for civil society enfranchisement and rural unrest. This line of thinking refutes the predominant western view that there has been an absence of political reform in China, and that the Chinese political system remains a stagnant and "ossified Leninist state" (Shambaugh 2008: 2). Shambaugh (2008: 3) considers this inaccurate view to be based on the presumption that if political reforms are not democratically-oriented, they are not valid and are automatically discounted as ineffectual. Failure to acknowledge the extensive range of proactive and incremental reforms that have been instituted by the party elite to consolidate its power and rule makes it difficult for outside observers to understand how the CCP has managed not only to survive but also become stronger, whereas so many other communist party-states have imploded. On this point Naughton (2010: 457) states, "It cannot be said that China has had no political reform; rather, the political reforms that have taken place have created a newly effective authoritarian hierarchy." The apparent effectiveness of reforms both to the inner-party, as well as within various sectors of the Chinese state, cannot be put down to luck. Shambaugh (2008) reveals how the Chinese leadership derived numerous lessons from the systematic comparative assessments of fallen communist party-states, as well as other communist and non-communist party-states in Asia, the Middle East, Europe and Latin America, that were undertaken with great urgency by the CCP following the Tiananmen protests and the demise of the Soviet bloc. The principal lesson that the party elite garnered from these extensive studies was "adapt and change – or atrophy and die" (Shambaugh 2008: 178). This realisation triggered efforts by the CCP to remould its ideological position to better accommodate current realities, and to restrengthen its organisational apparatus and capabilities. This demonstrated capacity for the CCP to learn and incrementally embrace a variety of new methods, of both foreign and Chinese provenance, is allegedly producing a robust political hybrid, which Shambaugh (2008: 181) terms China's "eclectic state".

Shambaugh's thesis persuasively refutes the dominant view that China's political system is unresponsive, hopelessly corrupt, fragile and in dire need of radical reform. Shambaugh presents a strong argument for the party's capacity to successfully adapt in order to meet an array of challenges. The record shows that the Chinese leadership has responded reasonably effectively to a broad range of social, political and economic problems. The CCP's success on these fronts is especially apparent when measured against the efforts of many other developing countries during the same time period. As a result political liberalisation is not on the agenda now or in the foreseeable future. The party's effectiveness in addressing internal and external challenges, along with maintaining social stability and economic growth, and facilitating the country's re-emergence as a great power, contributes to the political legitimacy of the regime, and arguably staves off demand for democratisation. The ability of the CCP to proactively learn, adapt and change is reiterated throughout Shambaugh's work. Similarly, Dickson (2003) discusses adaptive measures undertaken by the CCP in order to survive, notably by recruiting private-sector entrepreneurs as members (the 'red capitalists'), which is of course a significant departure from the party's traditional proletarian orientation. Dickson (2007: 244) suggests that as a result political power and economic wealth have become integrated in China. Such a strategy has not led to demands for democratically-oriented political reform as one might expect, "China's capitalists have not behaved as agents of political change. On the contrary, their very existence is the result of the party's policies, and their past and future success is dependent on the CCP's support. They have little interest in challenging the status quo that has allowed them to prosper, and they do not hold pro-democratic values that would lead them to press for fundamental change" (Dickson 2007: 243).

Thus far these evaluations of China's state capacity have focused on the internal scene, with a significant proportion preoccupied with the weaknesses of the regime. Experts who examine China's rise in world affairs often hold an entirely different viewpoint. For these analysts, China's impressive record of economic growth, rapid military modernisation, sophisticated foreign policy and authoritarian political system, are all considered to be indicative of a strong, coherent and effective state. China's ability to weather external shocks such as the Asian Financial Crisis in 1997 and the global financial crisis (GFC) in 2008, has also led many outside observers to believe that the Chinese state must be relatively strong and stable. Although not directly concerned with China's state capacity, the state's policy apparatus is nonetheless assumed to be centralised and coherent, with the core leadership able to systematically formulate and implement long-term strategies. Within this literature there have been few attempts to critically analyse China's political regime, especially in terms of the governance challenges it faces, with the notable exception of Shirk's *China: Fragile Superpower* (2007). Lampton (2005: 75) considers this one-sided approach in examining China's international behaviour, which has been adopted by many western scholars and policymakers, to be problematic stating, "One of the things that most worry Chinese leaders is that the strong-China paradigm makes it easy for foreigners to lose sight of China's genuine problems." The strong China paradigm has also given rise to the China threat theory, as many analysts worry about how China will use its newly acquired power on the world stage. The authoritarian nature of the regime obviously feeds into this fear factor over China's motivations and intentions abroad (see, for instance, Bernstein and Munro 1997 and 1998; Ross 1997; Gertz 2000). Naughton and Yang (2004: 5) sum up the nature of these polar characterisations of China as externally strong and internally weak, by saying that:

> Both raise important questions but do so in an alarmist and one-sided fashion. One sees only the problems and assumes the achievements are fragile; the other sees only the achievements and assumes the problems will be overcome. Neither takes adequate account of the extent to which the achievements and problems are intertwined, with the achievements possible only because certain problems could be left unaddressed, deferred to an uncertain future. Neither perspective seems to appreciate the extent to which its own conclusions should be modified by the insights of the other perspective. China is indisputably becoming a more diverse society, with a much larger economy, greater regional diversity, and many areas of social life slipping out of government control. At the same time, China has thus far managed to sustain national unity, and the government has proven itself remarkably resilient while its counterparts in many other transitional countries have fallen apart.

A large part of the problem with the contrasting characterisations that appear to dominate the extant literature is that they make little attempt to disaggregate the Chinese state, and thus fail to appreciate its hybrid nature, and the dramatic variations that exist in terms of state capabilities across policy sectors and geographies.

A more nuanced framework for understanding state capacity and institutional change during the Reform Era is provided by Naughton (2008a) who identifies two distinct phases in the process of China's market transition. In the first phase, 1978–1992, the focus of reform was on encouraging economic dynamism through incremental privatisation and decentralisation of state control, granting local governments the ability to experiment with new ways to achieve economic growth and development. From 1993 to the present day, the second phase of reform has been characterised by institutional reform and capacity building, namely recentralisation of political authority and resources (Naughton 2008a: 101–103). On this note, Bell and Feng (2009) give an account of institutional change 'Chinese style' that provides strong counter-examples to how institutional change in China is traditionally conceptualised. They argue that rather than being constrained by extant institutional arrangements (particularly the mid-level bureaucracy emphasised in the FA model), powerful party-state leaders have the ability to implement institutional change throughout the Chinese state. These powerful agents are "not 'sticky insiders', but powerful outsiders" to the institutions in question and when "prompted by certain exigencies, appear capable of reaching in and altering institutional arrangements" (Bell and Feng 2009: 121). Their view suggests that FA and the notion of a 'weak centre' do not continue to be a particularly accurate description of the Chinese political system. Instead the political system is characterised by a clear hierarchy of political authority and "marked by a steep power gradient, power can be quickly mobilised to effect change if needed, despite a deep-rooted bureaucratic inertia" (Bell and Feng 2009: 135). This indicates that the central party-state has far greater capacity than is typically accounted for in the extant literature. The frameworks and insights provided by Naughton (2008a), and Bell and Feng (2009 and 2013) are explored further in the next chapter, as they help us understand how the key to explaining policy outcomes rests with the interplay of elite and bureaucratic power within the Chinese state.

3
The Interplay of Elite and Bureaucratic Power

Scholarly explanations of Chinese political economy make much of its allegedly fragmented and decentralised character, which is deemed to have diminished the capacity of the central party-state to formulate and implement coherent, coordinated and effective policies. More specifically, policy inertia is often cited as a key problem, the result of protracted consensus building among multiple bureaucracies with roughly equal authority. In addition, other deficiencies such as rent-seeking behaviour and low regulatory capacity apparently plague and pervert the policy process. Such features of the Chinese political system are considered to be the result of the ongoing transition, beginning in 1978, from a centrally planned to a socialist market economy. The reforms associated with this transition saw a shift in the policy process away from the realm of non-institutionalised party leadership and toward a bureaucratically structured governmental apparatus (Huang 2004: 32–33). In addition to growing bureaucratic authority, political resources were further decentralised with the implementation of fiscal and administrative reforms, which transferred significant economic and fiscal power from the centre to local governments.[1] These trends in the first Reform Era resulted in the fragmentation and decentralisation of political authority within the Chinese state below the apex of political authority (the Politburo Standing Committee). The power and reach of the CCP was also somewhat diminished during the 1980s as party functions were reduced and redefined, and government accountability was expanded by "creating checks and balances on the power of the CCP" (Huang 2011: 14). Throughout the 1990s the party-state reversed many of these more liberally-oriented political reforms.

In the process of explaining the nature of China's policymaking environment during the Reform Era, Lieberthal, Oksenberg and

Lampton developed a model, known as fragmented authoritarianism (FA), in the late 1980s (Lampton 1987; Lieberthal and Oksenberg 1988; Lieberthal and Lampton 1992). At that particular stage of market transition the FA model was a fairly apt description of the policy process, especially with regard to economic decision-making. In contrast, policymaking in ideological policy areas such as media, education, culture and scientific research was always much less fragmented and diffused (Zhao 1995: 238–239). The FA model successfully captured the shift in the first Reform Era from "bureaucratic central command" through the state planning apparatus to interagency bargaining (Hamrin and Zhao 1995: xxvii). FA provided a compelling explanation for the growing weakness of the central party-state brought about by the processes of decentralisation and fragmentation of political authority that emerged with the onset of the Reform Era in 1978. Contemporary research on a wide range of policy issues and sectors in China continues to be informed by this model. Generally those who continue to apply this model harbour the assumption that the political and institutional landscape in China has failed to undergo any effective reform efforts towards reconfiguring these fragmented and decentralised arrangements. Such studies essentially perceive a paralysed political system 'standing off to one side' whilst everything else in China changes rapidly (Shambaugh 2008: 2–3). Significantly, these conventional accounts of the Chinese policy process also either downplay or ignore the role of the CCP as an institution and significant actor in the policy process, and instead privilege the role of bureaucratic and administrative institutions.

A number of scholars, notably Naughton (2008a), Shambaugh (2008), Pearson (2007), and Bell and Feng (2009 and 2013), now contest this conventional view. In his study of the Reform Era (1978–present) of Chinese political and economic development, Naughton counters the notion that transition strategies and the nature of reform policymaking have remained relatively unchanged throughout this period. He identifies two phases of the Reform Era characterised by distinct political economies and policy outcomes. According to Naughton (2008a: 101–102), a 'decentralisation of power and sharing of resources' characterised the initial period (1978–1992), during which time policymaking was highly incremental and power at the top was fragmented among multiple veto players. Reform during this time was experimental, *ad hoc* and encouraged local initiative, but was not couched within a coherent, long-term national reform strategy. Deng's axiom "crossing the river by feeling the stones" exemplifies the approach taken in the first Reform Era. At the same time, the central party-state "always reserved, and regularly exercised, the power to annul local experiments or to make them into a national model" (Heilman and Perry 2011). The second phase (1993–present) saw dramatic change in transition strategies with a focus on the recentralisation of political authority and resources; decision-making at the top became much more decisive, less fragmented and able to 'impose reform-related costs on specific social groups' (Naughton 2008a: 91–92). This study draws upon Naughton's useful distinction between the first and second Reform Eras, arguing that Chinese oil industry decision-making and development has become increasingly coordinated and effective during the second Reform Era. The range and extent of political reforms undertaken by the central party-state in order to reinforce its political authority is investigated further by Shambaugh (2008: 9), who shows that the flexibility, adaptability and learning capacity of the CCP is reasonably high (a feature uncommon to authoritarian regimes) accounting for its remarkable resilience within a globalised and liberalised world order, where most other communist party-state regimes have collapsed. Bell and Feng (2009: 121) offer an account of institutional change in China that rejects the assumptions of the FA model, particularly regarding the constraining effects of institutions on the policy process. Instead they claim that institutional change can be successfully implemented by powerful political elites (Bell and Feng 2009: 122). Hence the central party-state shapes institutional arrangements and policy outcomes rather than bureaucratic structure determining the policy process.

Given the changing conditions of China's political economy, there are other models that better describe the nature of political authority. One such model is BA, which emphasises the significance of elite power to the policy process, where governance is primarily a top-down, elite-driven phenomenon. This theory accounts for the recentralisation of political resources and authority, and also succeeds in bringing the party back into an analysis of the Chinese political system. The continuing marginalisation of the party and elevation of the bureaucracy and administration as defining features of the policy process in China fails to accurately capture the nature of the Chinese political system, which is characterised by a dynamic interplay of elite and bureaucratic power. What this chapter aims to show is that there has indeed been an overarching trend toward the recentralisation and reconstitution of hierarchical authority across many policy sectors, and especially within the strategic sectors or commanding heights of the Chinese economy. The reform policymaking environment has changed substantially and, in the case of the oil sector, cannot be simply characterised as decentralised and fragmented. The CCP remains in the driver's seat when it comes to economic reform and has made careful strategic choices about where to loosen its grip over Chinese society and economy, and conversely where to tighten it. Hence recentralisation does not appear to be a uniform phenomenon across the Chinese economy and administration, but rather than being indicative of a 'weak centre', this is more likely a reflection of conscious decisions enacted by the CCP.

This view does not fit with either the standard international relations (IR) or governance/policymaking explanations of the Chinese state, the former positing a unitary and monolithic political system, and the latter perceiving a weak and fragmented state. Rather, the understanding of the Chinese political system that is presented here is based on a sectoral approach that acknowledges more broadly the influence of a strong, capable and adaptable central party-state that retains key levers of authority and control throughout the political system and economy. There is a high degree of variance in the extent to which these mechanisms are utilised in order to solicit compliance, and is dependent upon the sector under investigation. Already mentioned, the strategic sectors of the Chinese economy are those that tend to command the greatest interest, control and investment on the part of China's political elites. These sectors have been the target of significant reform efforts over the past two decades for the purpose of recentralising authority and improving economic governance through the adoption of a regulatory state model, albeit one that is greatly distorted by various party-state institutions (Pearson 2007). Such efforts included streamlining relevant bureaucracies to enhance their capacity and efficiency, improving corporate governance in SOEs, strategically relinquishing smaller SOEs whilst retaining the larger ones and establishing regulators as market 'referees' (a partial solution to the problems arising from the absence of free market competition within state-owned sectors) (Pearson 2007: 718). That being said, the Chinese state is far from being monolithic and fully centralised. The heterogeneous nature of political authority and control across various policy sectors in China has been deliberately cultivated as a part of broader strategies for economic development formulated by the party-state elites.

The FA perspective on China's energy policymaking

Lampton (1987), Lieberthal and Oksenberg's (1988) influential studies of the Chinese political system during the late 1980s were among the first to downplay the significance of both elite politics and rational decision-making models, arguing that formal bureaucratic structures shape policy processes and outcomes. Lieberthal (1992: 8) states, "The fragmented authoritarianism model argues that authority below the very peak of the Chinese political system is fragmented and disjointed. The fragmentation is structurally based and has been enhanced by reform policies regarding procedures. The fragmentation, moreover, grew increasingly pronounced under the reforms beginning in the late 1970s..." and goes on to say, "Structurally, China's bureaucratic ranking system combines with the functional division of authority among various bureaucracies to produce a situation in which it is often necessary to achieve agreement among an array of bodies, where no single body has authority over the others." Based on these observations the FA model was developed in order to explain economic decision-making (as distinct from other policymaking spheres such as military, propaganda and so on) in China. The principal findings of the FA research agenda revealed a "fragmented bureaucratic structure of authority, decision making in which consensus building is central, and a policy process that is protracted, disjointed and incremental" (Lieberthal and Oksenberg 1988: 22). This institutional approach contrasts with the focus adopted by other scholars around this time, such as MacFarquhar (1981), Dittmer (1987) and Goldstein (1991), on informal political power dynamics within the party leadership, which were deemed to be highly personalistic. The focus of their studies was on policy disputes, factional divisions and power contests among the top leaders (Zhao 1995: 233).

Rather than being primarily top-down and hierarchical, Lieberthal and Oksenberg (1988) demonstrated that below the apex of China's political power structure authority was decentralised, fragmented and disjointed (although Naughton (2008a: 101–102) makes the observation that during the first Reform Era even power at the top was also fragmented among a group of "revolutionary elders" who wielded effective veto power, and derived their legitimacy and authority from their "personal histories, prestige and patronage networks"). This fragmentation was structurally-based and facilitated by the reform process during the 1980s, which saw the deliberate devolution and decentralisation of political authority to support market-oriented reforms drastically change the distribution of decision-making power in China. Hence in terms of the dynamics of economic decision-making, there was a marked shift from bureaucratic central command through the state planning apparatus to interagency bargaining. At the same time there was also a decline in the use of ideology as an instrument of control, and a reduction in the use of coercion, such as purges, against those who propose ideas that are eventually rejected, "thus emboldening participants to argue forcefully for their proposals" (Lieberthal 1992: 9). This loosened system and dispersed nature of political authority necessitated extensive horizontal bargaining and consensus building among numerous bureaucratic bodies of roughly equal power and authority (Lieberthal 1992: 9). In sum, the FA model essentially argued that organisational processes and bureaucratic politics shaped policy decisions and outcomes. Elite policy agendas could be readily altered and distorted at both formulation and implementation stages by the demands of this fragmented and protracted policy process (Lieberthal and Oksenberg 1988: 22), indicating that the flow of political authority within the system was in fact bottom- or middle-up rather than top-down.

FA has been the most important and enduring model in Chinese policy studies of the past two and a half decades, and was certainly useful in emphasising the significant role of the bureaucracy during this time. However, it tends to overstate the weakness of the centre and elite power in Chinese politics, and ignores the central party-state's carefully considered strategy of voluntarily relinquishing authority in some bureaucracies, typically those engaging in low politics, whilst restrengthening it in others (Hamrin and Zhao 1995: xxvii). Hamrin and Zhao (1995: xxvii) thought this model also erroneously implied an 'uncontrolled process' where mid-level bureaucratic actors have been able to seize autonomy against the will of the central party-state. This ignores the party-state's complicity in its own devolution, which has not at all been an even process across economic and industrial sectors. The model not only sidelines elite politics in general, but also ignores the role of the CCP as a powerful institution that is intertwined with state institutions. The party continues to wield significant levers of control within the state apparatus (many of which have been strengthened during the second Reform Era), such as the nomenklatura system and informal leading small groups and party committees (entities that do not appear on official organisational charts), in order to impose its particular policy preferences throughout the state sector.

Furthermore, Naughton (1992) and Halpern (1992) note that the two fundamental resources that structure bargaining positions tend to be in control over information and skills, and control over resources. The FA model focused more on the latter, failing to acknowledge and examine the importance of information to effective decision-making (Lieberthal 1992: 12). At the same time as reform efforts moved control over resources to lower bureaucratic levels, greater centralised control over economic information was created (Halpern 1992: 126). Halpern (1992: 131) identifies the important influence of policy research institutes created within the post-Mao bureaucracy, and concludes that they actually enhanced the capacity of the party leadership to undertake sophisticated long-term economic planning based on more comprehensive and accurate information. This example demonstrates that the FA model failed to capture some other significant forces of centralisation and capacity building occurring elsewhere in the party-state. That being said, Lieberthal (1992) clearly stated the FA model's focus on economic decision-making, as it was the policy sphere in which outside researchers could gain the greatest amount of access to information. Lastly, the FA model neglects other significant motivations for political behaviour that counter the constraints imposed by a fragmented bureaucratic structure. For instance, Chinese political culture contains strong centripetal forces that have influenced the development of the political system. Zheng (2007: 27) claims that a major enduring theme of Chinese political culture as it pertains to national integration is 'a great systemic whole'. This is a political value that has been maintained from the Chunqiu period through to the PRC (Zheng 2007: 27). Zheng (2007: 27) states, "Though the rise of regional forces often led to the breakdown of the Chinese Empire, the aim of all local power pretenders was always to restore the imperial order, a political value that was embedded in the popular saying 'striving to gain the political order of the Central realm' (*zhulu zhongyuan*)." This study does not delve into cultural or ideational motivations, but it is important to recognise that aspects of Chinese political culture allegedly enhance the coherence of their political system, hence countering fragmentation. The continuing relevance of the FA model should at least be questioned, as extensive reform efforts have been undertaken since the early 1990s, in order to recentralise political authority and control over resources. Since that time Chinese political economy has undergone substantial progress and change.

In *Bureaucracy, Politics, and Decision Making in Post-Mao China* (1992: 28) Lieberthal concedes that the CCP is a significant actor in the political system, but claims that its impacts cannot easily be observed and analysed due to the difficulty in gaining access to this information. Nonetheless, the lack of attention devoted to political elites is a major flaw of the theory, since arguably the CCP is the main actor that provides coherence and cohesion as a counter to the FA that these scholars describe. Despite Lieberthal's claim that CCP influence is difficult to analyse, the specific mechanisms that enable party control of the state apparatus are well known and include the nomenklatura system of personnel appointments and the use of leading small groups and party committees to oversee bureaucracies. In addition to acknowledging the significance of the party in the policy process, Lieberthal (1992) also claims that rational problem solving and personalistic politics continue to characterise the upper echelons of political and economic decision-making in China. Hence it is clear that scholars are aware of the value of these other approaches to policymaking in China, which are not factored into the FA model. However, they have not conceptualised the relationship among the various approaches or the circumstances in which each approach may prove valid. Zhao (1995: 233) notes, "This confusion reflects the actual ambiguous relationship between the roles of informal personal power and of formal institutional authority." Part of the problem arises from attempting to apply theories of policymaking in China to explain economic decision-making in general terms. This approach inevitably encounters problems as variations in the relative weight of elite and institutional authority in the policy process flow from different issue areas again, leading one to conclude that a sectoral approach is more appropriate (Zhao 1995: 238).

Studies of the energy policymaking framework in China still tend to rely upon this particular view of the country's bureaucratic tradition, where the majority of scholars have claimed that a fragmented energy governance structure has reduced China's capacity to produce robust, effective and responsive policies in a variety of energy sectors. For example, in their attempt to explain energy policymaking in China, and with reference to Lieberthal and Oksenberg (1988), Meidan et al. (2009: 597) claim:

> Policy initiatives are circulated amongst the different stakeholders for approval, allowing them to amend the drafts according to their interests. This procedure means that approval times are often lengthy and that the final policy proposal is a watered down version of the initial drafts. Furthermore ministries intervene at the implementation stage and have the power to stall or promote projects according to their interests.

Indeed Downs explicitly states, "The fragmented authoritarianism model generally explains Chinese energy security decision-making", which is seemingly exemplified by the lack of an energy ministry (up until 2010) to coordinate energy policies among the relevant bureaucratic agencies (Downs 2004a: 30). Hence conventional accounts claim that policy change with regard to energy is very difficult for China's decision-makers to achieve due to the limitations imposed by the institutional framework. This alleged incapacity of the central party-state is also clearly stated by Lester and Steinfeld (2007: 35):

> The real problem in China today, and the most important driver of the nation's energy and environmental footprint, is not geostrategic ambition, but rather a glaring deficit of governmental regulatory and administrative capacity...the real problem, overshadowing all others and least recognized by outsiders, pertains to the Chinese system's inability to govern coherently.

Similarly, Kong (2006: 64) goes so far as to identify the primary source of China's energy security as institutional insecurity, in terms of the inability of the country's energy institutions to meet energy challenges of both foreign and domestic provenance. In addition Kong (2005: 56) claims:

> China's risk aversion and poor energy policymaking system further magnifies its perceptions of low availability, reliability and affordability of oil imports, which further compounds its sense of energy insecurity...[China] lacks not only an energy policymaking system that can make and implement sound energy policies, but also an energy market that relies on market prices to allocate energy resources efficiently. As a result of this domestic failure, China has pushed its national flagship energy companies to undertake a global scavenger hunt for energy while muddling along a messy road of energy reform at home.

The decentralisation and fragmentation of the oil sector in China is also considered to have enhanced the political and economic power of the NOCs, allowing them to function as powerful independent actors in the Chinese political system (Kong 2010: 19–22). Downs (2008a), Houser (2008) and Kong (2010) all broadly claim that China's NOCs are increasingly independent of the party-state, comprising a particularly powerful interest group, able to influence political elites and shape oil industry development. The 'tail wagging the dog' aphorism is often invoked to characterise this dynamic.

The question remains as to why so many scholars persist in using the FA model when attempting to explain oil policymaking in China, without at least interrogating its continuing relevance given it was developed over twenty-five years ago. Perhaps this model remains the popular choice in Chinese policy studies among those who fail to acknowledge the significant state capacity-building efforts that have occurred particularly during the second reform period. Hence those who assume that the fundamentals of Chinese political system have not changed since the 1980s may uncritically adopt the FA model. Failure to acknowledge the party leadership's preference for a gradualist or incremental reform agenda might also account for the continuing popularity of this model. In this view scholars assume that a slower pace of institutional change is indicative of weak state capacity resulting from fragmentation, rather than it being the product of a deliberate reform strategy (this argument is discussed further in Chapters 5 and 6).

The neglected role of the CCP

At the same time as reform efforts have been geared toward improving state capacities during the second Reform Era, there has been significant intra-party ideological and organisational reform aimed at strengthening CCP discipline, effectiveness and legitimacy. In this study, the CCP as a distinct political institution rather than the political system 'writ large', is central to understanding state capacity and the dynamics of institution building in China (Shambaugh 2008: 1). The significant role of the party has been largely ignored, or else downplayed, both theoretically and empirically in the extant literature on Chinese policymaking. Since the late 1970s the Western literature on the Chinese policy process has typically entailed a close examination of state structures such as the administration and bureaucracy, with the role of the party and elite power relations being largely neglected (Brødsgaard and Zheng 2004: 3). During the 1950s, 1960s and 1970s the CCP occupied the focal point of research due to the fact that it was the sole actor in charge of nation-state building and socioeconomic development (Brødsgaard and Zheng 2004: 3). However, from the late 1970s new approaches to the study of Chinese politics emerged, which shifted the research focus away from the CCP. Brødsgaard and Zheng (2004: 5) claim the lack of attention afforded to the party politics in China partially reflects more general shifts in Political Science; since the 1970s studies of political parties have been marginalised, and in the 1980s 'the state' (comprised of the administration, bureaucracy and technocratic personnel) became the dominant unit of analysis. This was especially the case in studies on East Asian development, where scholars such as Wade (1990) and Johnson (1982) analysed the emergence of developmental states based on effective bureaucratic rule. Furthermore, the decentralisation of political authority that occurred during the first Reform Era also led many China scholars to neglect the influence of top-down mechanisms, such as the party (Holbig 2009: 38). That being said, during the era of Hu Yaobang and Zhao Ziyang (the most liberal Chinese leaders), which lasted from 1979 to 1988, attempts were made to redefine and reduce the functions of the CCP, and enhance the accountability of the government (Huang 2011: 14).

In the wake of the Tiananmen crackdown in 1989 some research interest among Western scholars was redirected to China's nascent civil society and other non-governmental developments, including the emergence of private actors resulting from the open door policy, but again with little or no consideration granted to the role of the CCP. In fact the focus of China studies has increasingly been on the "centrifugal forces in Chinese society", such as civil society, private-sector development, migrants and other marginalised groups "rather than the forces that hold the system together and make it work" (Brødsgaard and Zheng 2004: 2). Significantly, the marginalisation of the party in the relevant academic literature is also a result of the widespread belief that the CCP is obsolete, and has "difficulty surviving in a globalised world" (Brødsgaard and Zheng 2004: 5). The endgame of this trend toward obsolescence is typically predicted to be party collapse, resulting in a new political order characterised by "democratisation and power-sharing among different social forces and groups" (Brødsgaard and Zheng 2004: 5). However, many of these post-Tiananmen predictions of democratic transition in China have failed to materialise, and now scholars must grapple with the reality that the CCP has succeeded not only in maintaining its pre-eminent role in Chinese politics, but has managed to restrengthen and recentralise its political authority, which had dispersed and weakened to some extent during the first Reform Era.

The CCP comprises the core of power in the Chinese political system. This situation has not altered substantially since 1949, though at certain points in its history the party has weakened and faced major crises. Two instances where the CCP faced imminent collapse occurred during the Cultural Revolution, and the events of 1989–1991. However, in the wake of each disastrous episode the CCP managed to regroup, reform and re-emerge stronger than ever. The impact of the 1989–1991 period is relevant to this study as it directly shaped the

CCP's attitudes toward governance and policymaking in contemporary China, especially in terms of the need for the party to learn and adapt in order to address various socioeconomic challenges. With reference to the turbulence of 1989 Shambaugh (2008: 42) writes:

> During April and June 1989, the CCP, and China itself, was shaken to its core by the unprecedented popular demonstrations across the country, the violent suppression of them in Beijing on June 3rd and 4th, and the subsequent six months of martial law in the capital. Not since the Cultural Revolution had the CCP come so close to collapse. The party leadership was paralysed and split at the very top (resulting in the purge of the reformist CCP general secretary, Zhao Ziyang). The leaders felt embattled and believed that their rule, and the People's Republic itself, were at genuine risk of collapse.

Within six months of the Tiananmen massacre the unimaginable occurred – communist regimes elsewhere began to fall, culminating in the dissolution of the Soviet Union in 1991. The collapse of 'the birthplace of Bolshevism' in particular shocked the CCP leadership and 'provided cause for considerable introspection among its rank and file' (Shambaugh 2008: 53). In the wake of these events the CCP embarked upon a comprehensive range of studies of communist and non-communist party-states. The party leadership was particularly interested in identifying the causes of the implosion of communist party-states in the Eastern bloc and the Soviet Union. Instead of waiting for an inevitable collapse of power, the CCP decided that such a fate could be avoided through various proactive measures involving "introspection, adaptation and the implementation of pre-emptive reforms and policies" (Shambaugh 2008: 3).

Strengthening CCP authority and legitimacy through intra-party reform

China's post-Tiananmen efforts to understand the causes of communist party-state collapse, which were followed by extensive proactive efforts to implement appropriate reforms for the purpose of strengthening the party's ruling capacity, is indicative of the adaptability, flexibility and innovative capacity of the CCP. Whilst there were a host of very specific lessons coming out of the meticulous and far-reaching studies conducted by the party, the general lesson taken to heart by the CCP was that collapse occurred where the system of governance had become rigid (Shambaugh 2008: 126). Hence "While reacting to the events in former communist party-states, the CCP has been very proactive in instituting reforms within itself and within China" (Shambaugh 2008: 2). These reforms were intended to strengthen the party's ruling capacity, as well as the capacity of state institutions, and have been broad in scope. Again, this is an example of elite-driven institutional change, as the Chinese leadership has purposefully engaged in incremental expansion of central party-state capacities, while at the same time deepening market reforms. The CCP also developed an understanding that its own survival could not simply rely upon legitimacy derived from performance-based goals such as economic growth, social stability, governing capacity and accountability, as this basis for legitimacy was perceived within Chinese scholarly and policymaking circles to be fragile and unsustainable (Zhu 2011: 123). This is because performance-based legitimacy requires the government to fulfil those promises, so if Beijing fails to deliver on concrete promises the party then loses its basis for legitimacy, triggering political crisis (Zhao 2009: 416). Debates surrounding the 'performance dilemma' of party rule within China have given rise to recommendations on the revival, adaptation and innovation of party ideology (Holbig 2009: 38). Given that ideological communism as a system of governance lost credibility in post-Mao China, ideological legitimacy was effectively supplanted by performance-based legitimacy (Zhao 2009; Zhu 2011). In more recent years two factors in particular have contributed to the decline of legitimation capacity in China: widening inequality and rampant corruption (Wang 1995: 108). In response to this China's leaders have attempted to rebuild the party's ideological and moral legitimacy in order to provide a more stable and lasting justification for CCP rule. Under the Hu-Wen government in particular, intra-party ideological and organisational reforms were fundamental components of the CCP's self-strengthening movement.

The need to modernise Marxist ideology and improve the discipline and loyalty of the cadre corps were central themes in a key resolution adopted at the Fourth Plenary Session of the Sixteenth Central Committee in September 2004 entitled *The Decision of the Central Committee of the Chinese Communist Party on Strengthening the Party's Ability to Govern* (*China Daily* 2004b). According to this resolution the Chinese leadership's most urgent task was to enhance the CCP's ruling status, which could not simply be taken for granted (Lam 2006: 249). An intense rectification campaign aimed at "maintaining the advanced nature of CCP members" (*baochi dangyuan de xianjianxing*) followed the resolution and was implemented throughout the party-state apparatus over an eighteen month period (from 2005–2006) in order to reinstall party discipline and loyalty (Shambaugh 2008: 128–129). In order to enhance the Party's supremacy the resolution indicated "the party should strengthen leadership over legislative work", and also enhance its ability to exert further control over information and the media (Lam 2006: 250–251). Of course party domination of legislative and judicial work runs counter to notions such as the "rule of law" and "respect for the Constitution", which have also been touted elsewhere as ideals to which China should aspire (Lam 2006: 250). Shambaugh (2008: 110) and others have observed major crackdowns on dissent, the Internet and other potential challenges to CCP rule in recent years, and especially in the wake of the party-wide indoctrination campaign. This shows that party imperatives continue to trump other social, political and economic goals, which indicates that the CCP should indeed occupy a central role in research on decision-making and the policy process in China. In terms of how the CCP governs and maintains control in China throughout the state apparatus, there are two key dimensions to be briefly examined: ideology and organisation.

In the ideological sphere the adaptation of Marxist theory was considered a means to counter the fragility of performance-based legitimacy. The importance of ideology in legitimating party rule in contemporary China has been underestimated, as Holbig states, "the Chinese party-state, in its quest to legitimise authoritarian rule, has invested heavily in the continuous adaptation of official ideology to a changing domestic and international environment" (Holbig 2013: 76). Far from being rendered obsolete, the party's ideological foundations for legitimacy were re-emphasised during the Hu Jintao era of Chinese politics. That being said, ideology in China today has taken on a different meaning – it is no longer about adherence to abstract communist programs and principles, but rather is evaluated according to the party's capacity to respond to real social and economic challenges (Holbig 2009: 45). In other words, ideology has become more realistic and pragmatic; "a practical means to satisfy people's actual needs under the conditions of social transformation" (Lam 2006: 45). In today's China the crucial challenge for CCP ideology is to provide greater social equity and sustainable development. This challenge gave rise to Hu's theoretical contribution to party ideology, known as the Scientific Development Concept (*kexue fazhan guan*). Hu described this approach in a speech to the Central Party School in 2004 as meaning "economic and social development that are comprehensive, well-coordinated, and sustainable", also taking into consideration welfare and quality of life (Lam 2006: 42). The Scientific Development Concept also reaffirms the party-state's monopoly on power and position as the sole authority capable of assuming a redistributive role in the pursuit of social justice, which supports "the normative justification of its leading position in the country's modernisation process" (Holbig 2009: 49). This modernised ideological framework has been used to revive party legitimacy and guide policy. The Scientific Development Concept is frequently referred to in documents regarding China's energy policies, but whether or not ideology has a direct impact upon policymaking is not dealt with in this study. Rather it is important to note that the ideological revival has been undertaken to enhance party loyalty and the legitimacy of the regime. It reaffirms that the party pursues policy imperatives to ensure its own survival, and thus should certainly be considered when thinking about the policy process in China.

Another key finding concerning the reasons for communist party collapses in Eastern Europe and the Soviet Union was that these party-states had allowed their organisational capacity to deteriorate. In China the general direction of reform in the 1980s had been oriented towards economic decentralisation. This was reversed in the second Reform Era, as the party leadership turned its attention towards building an integrated state apparatus to support a market-oriented economy. In doing so, key levers of party-state control of Chinese economy and society were strengthened, the most significant being the nomenklatura system. This system determines which party members are eligible to fill top jobs in government, industry, universities and elsewhere. It allows the party to control "the appointments, transfer, promotion and removal of practically all but the lowest ranking officials" (McGregor 2010: 78). The secretive Central Organisation Department (COD) of the CCP is the official body that controls personnel appointments within the party-state. In the 1980s this department exercised a 'lighter touch' with regard to personnel selection, especially in Chinese universities, but this was reversed in the post-Tiananmen era of Chinese politics (McGregor 2010: 80). McGregor (2010: 84) claims that the COD treats the top management of state firms as if they were 'apparatchiks', to be 'shifted around at will, whatever commercial conflicts might arise', which emphasises the central role of the CCP in managing key personnel. However, it should be noted that the COD has codified rules for appointment that are not simply based on party loyalty, but also upon considerations such as length of service and education levels, in addition to participation in mandatory classes at a party school every five years (McGregor 2010: 81). Thus it is clear that there has been a concerted effort on the part of the party leadership to develop a professional, competent and loyal cadre corps to enhance the party's ruling capacity.

As mentioned above, the strategic sectors of the Chinese economy continue to be subject to party-state intervention. For instance, the heads of SOEs and the CEOs and boards of directors in Chinese listed companies are still appointed by the party. Lam (2006: 267) states, "Although the heads of such firms have assumed Western-style titles such as CEO and CFO, they retain traditional party and administrative rankings. Some are members or alternate members of the CCP Central Committee." While political loyalty is obviously an important factor used to determine these appointments, company performances also play a crucial role. This distinguishes the current era of personnel appointments from previous years where ideological commitment and relationships with the top leaders were the primary channels for career mobility in industry, which contributed to the growth of a highly unprofitable, inefficient and uncompetitive state sector (Lam 2006: 267). In order to encourage better performance among SOEs, SASAC has strengthened managerial incentives by introducing monitoring systems and linking the salaries of SOE executives to firm performance. Although the party-state intervenes in company management, rather than leaving such decisions to the market players, it has developed a more sophisticated capacity in this area to ensure that major heads of SOEs are professional, competent and competitive. Thus the nomenklatura system continues to provide the communist party with a particularly effective and pervasive instrument of control over the economy, where political and corporate loyalties generally reinforce each other.

Another important mechanism used by the party to monitor and control state bureaucracies, is leading small groups. They are supra-ministerial bodies that cut across the bureaucratic structure in order to facilitate consensus and coordination among party, government and military bodies. The use of these leading small groups has increased in the second Reform Era. They are an essential instrument created by the party to try and overcome bureaucratic infighting and policy coordination problems. The use of party committees and party groups in government departments and SOEs, which had been scaled back during the 1980s by liberal reform-minded Zhao Ziyang, have also been revived more recently (Huang 2011: 14; McGregor 2010: 80). These groups do not appear on official organisational charts, but they aim to ensure that the party line is followed and government policies are implemented. Hence these party committees and party groups enhance the CCP's strategic control over government departments and state firms (Kong 2010: 25). The expansion of these entities has further enabled the central party-state to establish a more centralised hierarchical political system with the party remaining at its core. This brief examination of how the ideological and organisational dimensions of the CCP have strengthened during the second Reform Era, demonstrates that China's political elites should not be neglected in research conducted on policymaking, and state-market relations in China.

A case for bureaucratic authoritarianism

BA avoids some of the main deficiencies of the FA model, and is usefully applied as a top-down bureaucratic account of the policy process in China. Importantly, it emphasises that interorganisational bargaining does indeed occur within a *hierarchical* state system. The BA model was first developed by O'Donnell (1973) to analyse the conversion to authoritarianism in many Latin American countries, such as Argentina, Uruguay, Brazil and Chile in the 1970s. O'Donnell's BA model challenged modernisation theory's suggestion that socioeconomic modernisation and democracy go hand-in-hand. Hamrin and Zhao (1995: xxv) later adapted the BA model to the case of China in the Deng Xiaoping era, arguing that the Chinese political system remains a "command system in terms of the top-down flow of authority (however negotiable the contents and terms of command may be), and party leadership directives are the primary means of regulating (*zhi*) the whole system". Bell and Feng (2013: 115) also provide clear definition of BA in the Chinese context claiming, "Under this model, 'bureaucratic' refers to the increasing institutionalisation of interest articulation and policy formulation through bureaucratic entities, as opposed to the centrality of unfettered discretionary power vested in revolutionary party leaders in Mao's era, while 'authoritarianism', on the other hand, pinpoints the monistic leadership of the Party in the overall political system. The Party commands, controls and integrates all other political organisations and institutions in China." While it does not discount the impact of horizontal bargaining among bureaucratic institutions on policy articulation and implementation, BA acknowledges that ultimate political authority rests with the central party-state, and that this authority can be imposed effectively where the need arises. Hence the preoccupation of the BA model tends to be more with the vertical relationships between political elites and bureaucratic agencies in the centre. The centrality of political elites in driving institutional change and continuity during the Reform Era of politics and economic development is further explained by Bell and Feng (2009: 123):

> In the process of systemic transition, China's bureaucracy in general terms either does not have the incentive to push for reforms that will curtail its superior authority and privileges enjoyed in the old days of central planning, or...resists reforms that will endanger its vested interests in the partial reforms and transition itself that facilitated the state predation and crony plunder. Given the especially sticky bureaucracy in China's context, the party elites, as powerful external agents, found themselves being the drivers of reform/institutional change, seeing their legitimacy increasingly resting on economic performance that could only be achieved through continuous reforms.

Shirk (1992: 76) also highlights the ultimate authority of the political elites in arbitrating bureaucratic conflict and enforcing critical policy directives, through 'management by exception,' where you have "powerful agents who are external to given institutional environments" and can play a significant role in institutional change (Bell and Feng 2009: 118).

There is compelling evidence in support of the significance of elite power in shaping policy outcomes in China. Bureaucratic restructuring is a solid indicator of the top-down authority and ruling capacity of the CCP, especially since within the Chinese state institutional reorganisation is the only way to achieve power redistribution in the absence of democratic mechanisms (Zheng 2004: 84). If the central party-state is weak, then it is unlikely that bureaucratic interests and resistance can be overcome. However, since the early 1990s there have been several successful rounds of extensive bureaucratic restructuring undertaken in order to improve bureaucratic efficiency, meet the demands of an emerging market economy, and also to consolidate and strengthen the power of the central leadership (Zheng 2004: 84). While the conventional view that China's institutional environment imposes major constraints on decision-making can explain bureaucratic stasis and continuity, it cannot account for institutional change (Bell and Feng 2009: 120). Bell and Feng (2009: 122) show that political elites in China retain sufficient political authority to take "significant policy initiatives and direct intervention on its key concerns, and have the final say in balancing conflicting interests, ironing out strong disagreement in the bureaucracy and pushing through bold programs on tough issues or issues with "overwhelming urgency"." China's industrial policies towards the state sector during the 1990s and 2000s have also been markedly top-down in character, and have drawn little input from business communities and SOEs (Huang 2008: 280–281). Hence the BA model provides a more apt explanation of policymaking during the second Reform Era since elite power has become stronger and more cohesive, and the party-state has undergone a series of reforms aimed at recentralising the country's administrative and organisational apparatus.

The BA model goes further towards explaining the overarching interplay of elite and bureaucratic power within the Chinese state. However, it also has its flaws, as Bell and Feng (2013: 116) claim, BA runs the risk of shifting attention too far back to an elitist account, and also fails to provide a complete understanding of the evolving elite and bureaucratic relations during economic transition, where some key bureaucracies have become empowered to the point where they can resist or sometimes work around central directives, particularly those that have become more market-oriented such as the People's Bank of China (PBC). The main problem with both the FA and BA models is their underlying assumption that the political authority only flows in one direction, either bottom-up or top-down. This view of the interactions between the central party-state and various bureaucracies fails to capture the complexity of the relationship between elite and bureaucratic power. Whilst the flow of authority within the Chinese political system generally seems to be top-down and hierarchical, especially within the realm of high politics and the strategic policy sectors, certain bureaucracies can resist elite policy agendas or at least force compromise positions. This seems to indicate that in some cases political authority flows both ways, depending on the type of bureaucracy and policy issue involved, which is why sectoral studies are more appropriate for this kind of analysis. Certainly the central party-state may choose to delegate authority downwards in some cases where it is clearly beneficial to do so, for instance, in a study of the PBC Bell and Feng (2013: 137) argue that central banks "can *win* authority if they are able to set the agenda and establish a credible policy and administrative track record and adeptly utilise institutional resources and capacities." The Chinese leadership maintains ultimate authority over the strategic direction of all the bureaucracies within the party-state, but is willing to share 'parcels of authority and policymaking', especially where 'indispensable' knowledge and expertise exists within particular bureaucratic bodies (Bell and Feng 2013: 137).

In contrast to most extant research on China's oil sector, this study relies on a party-state centred approach, which is underpinned by the BA model (with the acknowledgement that although political authority is primarily an elite top-down phenomenon, there are significant instances of authority sharing and delegation downwards to some bureaucracies), in order to provide an explanation of China's response to various oil security challenges. The focus is not simply on horizontal interorganisational bargaining as emphasised by the FA model, but also involves an examination of elite power and the instruments of authority and control retained by party elites to implement their policy agendas. Most importantly, the BA model recognises the steep hierarchy of political authority that exists in the Chinese political system and the power of party elites to push through policy reform. Hence the relationship between the political elite and bureaucracy is considered a key variable in shaping policy approaches and outcomes. Rather than being perpetually hamstrung by the bureaucracy, political elites in China can indeed effect institutional change, and have done so when a given policy sector is sufficiently significant to draw attention from the Chinese leadership. Scholars such as Shambaugh (2008), Yang (2004) and Bell and Feng (2009 and 2013), provide some persuasive examples of elite-driven institutional change. Without an account of the role of party elites in the reform process institutional change in China cannot be readily explained. The following empirical chapters make the case for their BA account of institutional change in China's oil sector.

Note

1 The 'centre' is the national level of government and includes the State Council and its commissions, ministries, leading groups in Beijing, the Party Politburo, Secretariat and the organs of the Central Committee. 'Local' government is the provincial, municipal or county level of government.

4
The Socialist Era of Oil Self-Sufficiency (1949–1977)

During the PRC's early years the development of the oil sector was heavily influenced by revolutionary fervour, ideological campaigns and the nationwide drive for rapid industrialisation. The oil industry became a locus for these ideational and structural forces to the point where the Maoist innovations achieved within this sector became the model for national industrial development across all economic sectors. For the first thirty years of its history the PRC pursued a vision of socialism that entailed a centrally planned economic system, and the development of a massive socialist industrial complex through direct government command and control. Naughton (2007: 55) labels China's development strategy during this time 'Big Push industrialisation', because the overwhelming priority was to channel the maximum amount of resources and investment into heavy industry. Hence the Big Push strategy shaped virtually every aspect of the Chinese economy. The command economic model, based on the system created in the Soviet Union under Stalin, was adopted in order to implement this strategy. Under this system of economic governance the central party-state directly allocated resources, set production targets and controlled the pricing system. By the late 1950s Mao Zedong had begun to doubt the applicability of the Soviet model to the Chinese context and embarked on a program to accelerate economic development called the Great Leap Forward, which to a degree decentralised the economy. Throughout this entire period, the political imperatives of the central party-state, rather than market forces shaped the content of China's economic blueprint and this of course included the oil sector. Such an approach complemented Mao's strategy of national self-reliance and limited openness to the rest of the world,

The Socialist Era of Oil Self-Sufficiency (1949–1977) 71

which came to the fore in the wake of the Sino-Soviet rift beginning in 1959.

The development of the oil industry during the socialist era can be further divided into two distinct phases; (1) significant dependency on the Soviet Union (1949–1959) followed by (2) the drive for, and subsequent achievement of, energy independence under conditions of economic autarky (1959–1978). One of the most urgent problems facing the PRC after 1949 was energy development. The following decade would be characterised by a high degree of dependency on the Soviet Union for oil imports, and also expertise and technology to aid the development of China's nascent oil industry. Due to the extent of this dependency the Sino-Soviet rift dealt a severe economic blow to China. From then on Maoist principles of national self-reliance, self-sufficiency and economic autarky drove the rapid development of the country's oil sector. The 'Daqing method,' named after the China's most successful oil field, which became a symbol of modernity in China, emerged out of Soviet estrangement and also in response to Mao's desire to advance a model of socialist development more suitable to the Chinese context than Soviet methods. The new Maoist model of economic development stressed self-reliance, political education and the input of labour rather than reliance on scarce capital resources and the use of advanced technologies (Kambara and Howe 2007: 14). 'Learn from Daqing' became the catch-cry of China's industrialisation drive, as the Daqing oil field successfully drew upon class struggle "to promote a specifically Chinese model of development" (Fenby 2008: 459). Hence energy development in general and the Daqing project in particular, became the nation's symbol of political strength and its ability to survive without foreign support.

Due to its vital strategic importance, central planning and the command economic model ensured tight party-state control of China's oil sector. Nonetheless, some operational decentralisation occurred during the Great Leap Forward, which led to the adoption of a commune system approach to oil exploration and production. For example, the Daqing oil field was organised as if it were a large commune in which industry, agriculture, schools and even the people's militia played a role (Hama 1980: 186). The achievement of oil self-sufficiency through mass mobilisation was a product of extant resource distribution, and various state capacities and deficiencies, namely abundant labour and a shortage of capital. The state's capacity to control and direct oil sector development was high, and China achieved success early on when measured against the professed aims of oil sector development, which were primarily to support heavy industry and eliminate dependency on foreign sources of energy. China became self-sufficient in oil in the mid-1960s and an oil exporter in the early 1970s. By 1985 China exported over 25 per cent of its petroleum production and earned over 20 per cent of its foreign currency through oil exports (Lieberthal and Oksenberg 1988: 169). Since the development of the oil sector was regarded as one of only a handful of successful endeavours driven by Mao Zedong Thought, the key leaders within this sector, known as the petroleum group (*shiyou pai*), became one of the most powerful factions in the history of PRC politics. This group was able to influence broader economic policy objectives in the service of energy sector imperatives, albeit within the parameters established by Mao Zedong Thought. The existence of this interest group during the socialist era of oil industry development further enhanced capacity within this sector, as members of the faction had direct access to Premier Zhou Enlai, which allowed the petroleum group to advise and influence the party leadership.

Despite achieving oil self-sufficiency by the end of the 1960s, it cannot necessarily be said that China possessed energy security. This is because China's command economy was unable to allocate resources efficiently, and the overarching political and economic goals of industrial and military development were far too ambitious, to the point that by the mid-1970s the Chinese economy was on the verge of collapse (Zha 2006: 179–180). While the ideological preference for self-reliance and economic autarky produced some early successes, as time went on these approaches began to take their toll on energy supply as well, especially since China could not keep up with technological advances in oil exploration and production that were occurring elsewhere in the world. Once the readily accessible deposits of oil had been depleted, it became very difficult for China to maintain oil output on the back of mobilisation of the masses and increasingly out-dated and inefficient procurement methods. The socialist era of oil sector development provides a good example of strong state intervention combined with poor decision-making capacity, due to the highly politicised context where ideology rather than economic reality reigned supreme. From the early 1970s China's policy towards its energy resources started to evolve away from the commitment to self-reliance in response to problems that arose from the pursuit of this objective. Lieberthal and Oksenberg (1988: 171) claim that China's leaders "increasingly perceived two overpowering realities: foreign equipment, technology, and capital would greatly facilitate the rapid and effective development of China's petroleum resources, especially offshore; and oil exports could help finance imports which would accelerate China's industrialization." However, it was not until 1978, with the consolidation of Deng Xiaoping's leadership and power, that petroleum policymakers were fully released from the exigencies of Maoism.

This chapter traces the development of China's oil industry mainly during the socialist pre-Reform Era. The purpose of this historical narrative is to show the Maoist system bequeathed a legacy of strong institutions, policy instruments, and organisational and political capacity to foster oil industry development. The developmental failures that began to appear by the 1970s, such as gross oil waste and inefficiency, and technological stagnation, were mainly the result of a radicalised political environment and the inefficiencies caused by central planning, rather than weak state capacity *per se*. Aspects of the institutional environment established during the Maoist era actually proved highly effective when reoriented to economic modernisation by Mao's successors.

Early lessons in the pitfalls of foreign oil dependency

China's oil industry remained small, underdeveloped and primarily located in the country's west up until the communist takeover in 1949. Chinese oil field exploration and production began around the turn of the twentieth century, when Qing government officials invited Japanese engineers to prospect for oil at Yanchang and Shanxi provinces (Kambara and Howe 2007: 7). Despite the abundant oil deposits located in those provinces, financial and technological constraints presented major limitations to their effective exploitation. Oil field developments up until 1949 were small and *ad hoc*, and were often undertaken with assistance from the Soviet Union and Japan during the earlier part of the twentieth century. China's domestic oil production remained insufficient to meet domestic demand and hence the country remained heavily dependent upon oil imports. When the CCP came to power in 1949 it inherited three domestic onshore oil fields – Yumen, Yanchang and Dushanzi (all located in the northwest) – that together had an aggregate output of 2,000 barrels per day (Ma 1980: 99). China's underdeveloped petroleum industry had been stimulated by the domestic oil exploration and production that was undertaken as a wartime measure during the Second World War, largely as a result of the Japanese blockade (complete Japanese occupation of the coastal ports in 1941 and the closing of the Burma Road), which cut off

China's oil imports (Smyth 1946: 187 and 189). However, there were two main obstacles that hindered China's domestic oil production efforts during the war: poor transportation and lack of modern technology (Smyth 1946: 189). Hence the problem of energy development was perceived to be a most pressing and urgent task after 1949. In April 1950, China's first National Petroleum Congress was convened and the Ministry of Fuel Industry was given overall responsibility for petroleum. In 1955 the Petroleum Administration Department within that Ministry was upgraded to the rank of Ministry of Petroleum Industry (MPI). The MPI initiated and supervised all activities relating to exploration, oil-field development, and the construction of refineries (Kambara and Howe 2007: 11). Following the communist takeover China became the largest importer of Soviet crude and oil products, and remained so until 1961 when it was surpassed by Czechoslovakia (Klinghoffer 1976: 540–541). China paid a high price for Soviet oil, and Klinghoffer (1976: 541) claims that this was to an extent "due to the overpricing of goods by both sides in an essentially barter trade, but some degree of exploitation was also present since the Soviets had largely captive markets in other communist-ruled states".

In 1949 Mao established a diplomatic alliance with Stalin, the primary aims of which were to deter possible aggression from Japan and the United States, and to gain Soviet assistance for China's economic reconstruction and industrialisation (Zhang 2001: 61). From 1950 to 1960 Soviet Russian specialists were sent to China to work on oil exploration projects in areas such as Daqing (although the first field discoveries were made after the Russians had left) and Xinjiang, where a joint venture called the Sino-Soviet Oil Company was created in 1950 to further develop the Dushanzi oil field. Technical dependence on the Russians went hand-in-hand with the adoption of Soviet-style administrative arrangements. Planning was centralised, and under the MPI an Oilfields Administration Department and Exploration Supervision Department were formed (Kambara and Howe 2007: 11). However, these accomplishments presented Mao and his comrades with an "inevitable paradox", as Zhang (2001: 61–62) notes:

> On the one hand, the CCP's aspiration for political independence would forbid the regime to rely too heavily on Soviet economic aid, because economic dependency sacrificed political leverage; on the other hand, the enormity of China's economic problems would compel the new regime to seek as much foreign aid as it could. How to achieve a balance between obtaining sufficient Soviet assistance

and relying too much on the Soviets proved to be one of the CCP's gravest challenges in sustaining the alliance.

Although Mao had hoped the Sino-Soviet alliance would be based on equality and solidarity, the Chinese soon began to sense that the Soviets were taking advantage of their country's difficult situation. For the Chinese, the Soviet Union's insistence on the establishment of seemingly exploitative joint ventures in China, and also Stalin's strict demands for special treatment for Soviet advisers were indicative of 'Russian chauvinism' (Zhang 2001: 62–68). This subjugation to Soviet authority would be relatively short-lived as China sought to regain its autonomy and independence in the late 1950s.

The Sino-Soviet split led to a phasing out of Soviet oil supply to China. The Soviet Union attempted to use oil as a political weapon in order to stem Chinese moves towards political autonomy, and coerce it into realigning with the Soviet-led communist bloc (Klinghoffer 1976: 542). Klinghoffer (1976: 542) claims that the Soviet Union pursued 'gradual and selective' application of economic warfare against China in order to keep the option for rapprochement open. For instance, after 1960 a Soviet embargo on crude oil was implemented, which was a tactical choice since China purchased mainly oil products rather than crude, due to limited domestic refining capacity (Klinghoffer 1976: 540, 542–543). However, deliveries of certain oil products, such as jet and aviation fuel, were cut back resulting in a reduction in Chinese military flights (Klinghoffer 1976: 543). Following Soviet and Chinese clashes at Communist Party meetings and congresses in Bucharest, Peking and Hanoi in 1960, Khrushchev abruptly withdrew Soviet technical personnel from China and all oil assistance programs were ended, leaving many major projects unfinished (Spence 1980: 285). Wang Qumin, one of Daqing's engineers, recalls the Russians' mocking taunts as they departed, "They said, 'Your methods are too primitive. Trying to develop Chinese oil fields without Russian expertise is as futile as trying to learn how to make a shirt by studying pieces of cloth'" (Chandler 2004: 116).

Soviet oil sales to China finally ended in 1969 as a result of the Sino-Soviet border conflict on the Ussuri River, which led to a further deterioration in political relations (Klinghoffer 1976: 544). The overall effects of Soviet actions impacted China materially and psychologically; they temporarily derailed China's economic development plan, arguably contributed to the economic crisis of 1960 to 1961, and also added strong impetus to China's desire to achieve strength through self-reliance and avoid further national humiliation at the hands of interfering foreign powers. In other words, this experience of dependence reinforced a tendency to nationalist and mercantilist policy approaches. In addition the Sino-Soviet rift also intensified a deep questioning by China over the entire Soviet approach to economic development and planning. The Soviet approach essentially embodied both a strategy for resource allocation, which gave priority to heavy industry, and a method of economic administration through the use of centralised bureaucratic planning (Kambara and Howe 2007: 14). Mao became convinced that these fundamentals were inapplicable to the case of China. Hence he turned industrial development, with a particular focus on the oil sector, towards a new 'Maoist' method of development, which emphasised self-reliance, political education and the input of labour (Kambara and Howe 2007: 14). In addition to the loss of Soviet assistance, the Paris-based Coordinating Committee for Multilateral Export Control (COCOM), comprised of Japan and all North Atlantic Treaty Organization (NATO) countries except Iceland, which controlled exports to communist countries, placed a ban on the export of advanced industrial equipment and technologies to China further hindering the country's oil sector development (Lim 2010: 77). Lim (2010: 78) notes that up until 1967 most of the challenges China faced in developing its domestic oil industry were generated externally. This would change from the mid to late 1960s onwards with the onset of the Cultural Revolution, and the emergence of energy problems arising from China's political and economic isolation.

The pursuit of self-reliance during the Great Leap Forward

Initially the socialist institutional landscape was centralised and tightly controlled by the central party-state. The CCP was the sole actor in charge of the country's development, and exerted pervasive control over all aspects of Chinese society and economy. Soviet-style administrative arrangements were adopted at the outset to implement central commands and directives. However, debates over China's growth strategy, which occurred during 1957 and 1958 in response to disappointing rates of agricultural production, saw China move away from Soviet-style planning and methods, and pursue greater decentralisation with regard to economic decision-making (Spence 1999: 545). Mao's aim was to stimulate agricultural production by adapting the institutional environment to China's specific needs, rather than continue to rely on a political and economic structure directly transplanted from the Soviet Union. Centralised leadership with dispersed operations was deemed to be appropriate to deal with China's widely distributed resources and labour. Hence Mao advocated the use of central planning agencies to "coordinate the production of the Party, government and army systems by drawing up overall indicators for such things as use of raw materials, production totals, and the distribution of products. Once the plan was decided it should be handed over to the separate systems for implementation" (Mao 1980: 36). Partial decentralisation occurred according to system and area, rather than sector. This was essentially a compromise between a fully centralised organisation that tends towards inflexibility and does not provide adequate incentives for initiative at lower levels, and complete decentralisation, which would result in poor coordination and inefficiency (Mao 1980: 36).

These ideas culminated in the Great Leap Forward (1958–1960), which was a strategy of dramatically expanding production through "the mobilisation of mass will and energy, especially when freed of the constraining effects of overcautious planning and an entrenched bureaucracy" (Spence 1999: 549). Local party leaders were expected to provide the inspiration and impetus for this mass mobilisation, with the goal of achieving a fully communist society. At the same time, the Big Push strategy of industrialisation was also intensified. Mao thought that by "decentralising economic decision making, this strategy would lead to even greater CCP power in the countryside and a corresponding decline in the influence of professional economic planners in the ministries" (Spence 1999: 549). Hence while political and ideological control remained exceptionally tight (even more so than in the Soviet Union), economic decision-making was decentralised to a certain extent. The Great Leap Forward was quickly discredited as it turned out to be a catastrophic failure, resulting in famine and near-economic collapse. However, there were a handful of areas where Mao's plan of establishing a commune system to promote mass mobilisation and self-reliance appeared to work, albeit superficially, one of which was the Daqing oil field.

The realisation of oil self-sufficiency under Mao

The proposed solution to the energy insecurity that resulted from the loss of Soviet assistance and economic collapse was to develop the Daqing oil field at a maximum rate and to do so without reliance on foreign technicians, capital and equipment (Lieberthal and Oksenberg 1988: 172). In the early part of the twentieth century China was believed to be deficient in oil reserves because "conventional Western geological knowledge" dictated that major oil deposits were formed in marine basins, which China did not possess (Ma 1980: 99). The Chinese geologist and first minister in charge of geological prospecting at the Ministry of Geology (MOG) Li Siguang rejected this "oil deficiency" thesis and argued that oil could be found in the "unique tectonic structures of the continental lacustrine basins of China." Li's research eventually led to the discovery and exploration of oil in the Daqing and Shengli regions, as he recommended to Mao that prospecting efforts should be transferred from western China to new potential fields in the east (Kambara and Howe 2007: 12). Strategically this was also an important discovery as China's heavy industries were mostly located on the east coast, which required oil to be transported over vast distances from the country's western interior (where China's main oil reserves were being exploited) (Lim 2010: 66). The fact that Chinese geological theories aided these major oil discoveries was also a source of pride in the context of Mao's economic development strategy stressing self-reliance and the eschewal of foreign knowledge systems (Lieberthal and Oksenberg 1988: 176). Since the timing of its rapid development was a partial result of both the scaling back of Soviet oil imports and assistance to China and the disastrous Great Leap Forward programs, which caused a fall in coal production (exacerbating an energy crisis), Daqing became the national symbol of self-reliance (*zili gensheng*) and Chinese-led industrialisation. Oil industry development thus became an integral part of China's economic development strategy.

Daqing was producing 88,600 barrels per day by 1963, which allowed Mao to officially declare China's energy independence (Ebel 2005: 7). This news was publicly announced in a summary communiqué of the Fourth Session of the Second National People's Congress (NPC), which was published in the *People's Daily* on December 4, 1963 (Lieberthal and Oksenberg 1988: 184). The drive for national self-reliance is a recurrent theme throughout Chinese history, and especially during the communist period. Post-1949 this desire was essentially motivated by nationalist pride and the traumatic historical experience of colonialism, which led to the formation of a view, widespread among both party-state elites and the general population, that the outside world wanted to prevent China from becoming a great power. Woodard (1980: xviii) argues that China's self-reliance policy developed in response to the "national humiliation, unification and revolution" following the country's encounters with the West. It was an essential part of China's struggle to regain its technological and scientific independence and pre-eminence (Woodard 1980: xviii). In December 1963 Mao announced, "Among the dozens of ministries under the central government there are obviously several which have done better and have a better style of work, for instance, the Ministry of Petroleum Industry (MPI). Yet the other ministries simply ignore them and have never bothered to visit them, study their experience and compare notes" (Hama 1980: 194). In 1964 Mao issued the Supreme Directive 'In industry, learn from Daqing', and during the 1960s and 1970s Daqing became the national industrial model for all other industrial sectors to emulate. Daqing was a main focus of Maoist propaganda campaigns that manufactured images and narratives of heroic workers selflessly striving to fulfil the requirements of Chinese-directed socialist modernisation. Daqing represented the most successful application of Mao Zedong Thought to the problem of industrialisation (Lieberthal and Oksenberg 1988: 178). The successes of the oil industry operating on Maoist lines resulted in the rise to political prominence of the petroleum group (*shiyou pai*), headed by Yu Qiuli and other individuals who helped to develop the Daqing oil field (Lieberthal and Oksenberg 1988: 46).

The 'big battle' formula, informed by Mao Zedong Thought, was adopted as the method for exploiting new oil resources. In February 1960 the Central Committee of the CCP launched a 'big battle for oil', and for that purpose assigned nearly 100,000 workers, military men, experts, and cadres to the project (Hama 1980: 185). This campaign was an attempt to apply Chinese military thinking to the development of the economy (Hama 1980: 185). The mobilisation of workers into a battle formation was considered a solution to the challenges presented by technology and capital shortages. To implement this strategy the State Council gave supreme priority to the Daqing project, requiring all relevant ministries (transport, machinery, construction, railroads, agriculture and forestry) as well as the Heilongjiang provincial government, to cooperate under the umbrella control of the MPI (Kambara and Howe 2007: 14). At any one time there were 40,000 workers mobilised in the Daqing oil fields (Kambara and Howe 2007: 14). In essence Daqing functioned as a large commune, where both industrial and agricultural work was encouraged in order to form a self-contained, self-reliant unit (Hama 1980: 194). Schools, hospitals and housing were also provided to enhance Daqing's capacity for self-sufficiency. In this context, the 'Iron Man' Wang Jinxi, a drilling crew chief in Daqing, became China's archetypal and pre-eminent 'model worker'.

Wang led the No.1205 drilling team, which struck Daqing's first production well (the Songji-3 well) on September 26, 1959, allegedly after five days of non-stop work. Wang Jinxi was noted for his strong work ethic, commitment to socialist reconstruction, and, according to communist legend, for a defining moment when he leaped into a slurry pit with a broken leg in order to stir the cement with his gyrating body (on account of a broken mixing pump) so it could be dumped on a gusher that was spewing oil into the air. (Kynge 2006: 128–129). Consistent with Mao's ideological commitments and anti-intellectualism the model workers were well drillers, and the more significant contributions of scientists and technicians to Daqing's initial success were downplayed (Lieberthal and Oksenberg 1988: 178–181). Premier Zhou Enlai (quoted in Hama 1980: 194) was particularly interested in the progress made at Daqing and stated in a report to the First Session of the NPC, delivered in 1964:

> Daqing is an example of the study and application of Mao Zedong Thought. Their slogan is "building the oil field on the basis of two theories," which means studying Chairman Mao's "On Practice" and "On Contradiction." They made use of the dialectical and historical-materialistic stand, views, and methods advanced in the two works to analyse, study, and solve all kinds of problems brought about during the construction. Daqing is also an example of learning from the People's Liberation Army and applying its experience in political work. Daqing all along adhered to the principle of combining centralized leadership with mass movements, the principle of combining high revolutionary spirit with a strict scientific approach, the principle of making technical revolution, and the principle of building the country through industry and frugality, and thus it fully met the requirements for greater, faster, better, and more economical results set by the general line for building socialism.

Ultimately this ideological dimension of China's economic development strategy had adverse affects on domestic oil production. In an effort to "win propaganda plaudits" by finding and extracting oil as quickly as possible, drilling teams pursued reckless and careless methods (Kynge 2006: 129). As a result enormous quantities of oil, up to 90 per cent according to Kynge (2006: 129), were wasted, either left unrecovered or spent in gushers. Furthermore, the purge of intellectual and technical elites during the Cultural Revolution, and Mao's pursuit of autarkic economic policies resulted in scientific and technological stagnation. During the 1960s Maoist policies also stressed human labour power as the most important component in the forces of production, with scientists and foreign knowledge systems being regarded with suspicion (Kerr 2007: 82). Hence old machinery and inadequate drilling techniques contributed to the inefficient exploitation of China's oil fields. Paradoxically the political measures that were introduced in order to pursue self-reliance through self-sufficiency actually served to exacerbate China's energy insecurity in the long run.

Construction of the third front and the Cultural Revolution

During the initial stages of China's oil sector development after 1949 the main energy challenges the country faced were generated externally. The severance of Soviet oil imports and assistance, combined with a western embargo on oil technology exports to China, drove rapid energy sector development and contributed to the preference for autarkic economic policies that revolved around self-sufficiency and self-reliance. The adoption of these policy approaches yielded developmental successes early on, and by 1963 China had become self-sufficient in oil. However, once the immediate imperatives of economic survival had been met, the party leadership began to turn its attention towards longer-term economic development plans, the objectives of which could no longer be met effectively through a strictly self-reliant policy stance. Furthermore, from 1964 the international political context began to change both for better and worse: Chinese relations with Japan and western Europe began to improve – Sino-Japanese unofficial trade resumed, and French president Charles de Gaulle granted official recognition to the PRC (Lieberthal and Oksenberg 1988: 181). At the same time, the risk of Sino-American confrontation, as a result of the United States' growing presence in Vietnam, began to increase (Lieberthal and Oksenberg 1988: 182). The Vietnam War prompted Mao to prepare for the possibility of a Sino-American war by creating the 'third front' (an industrial base in the southwest of the country, which would be strategically secure in the event of war). This third front represented the strategic and economic dimension of Mao's anti-Soviet and anti-American policies (Kambara and Howe 2007: 25).

Mao was concerned that the concentration of factories along the country's coastal regions gave rise to geostrategic vulnerability, and argued that the industrial development should be shifted to the interior. In order to achieve rapid development of China's inland industrial base, Mao established a new office in 1964 – 'national economy supreme command', otherwise known as the 'small planning commission', which granted the centre an even higher degree of concentrated authority over economic planning (Lieberthal and Oksenberg 1988: 188). This move was a response to Mao's dissatisfaction with the State Planning Commission's (SPC) preliminary thinking on the Third Five-Year Plan (1966–1970), where the commission's economic planners essentially ignored Mao's desire to develop the third front (Lieberthal and Oksenberg 1988: 189). MPI minister Yu Qiuli played a lead role in this new body, and assisted in the development of the third front plan (Lieberthal and Oksenberg 1988: 189). The oil sector continued to play a lead part in industrial development, due not only to its crucial role in energy supply and as a model for other industrial sectors to follow, but also as a critical element in Mao's third front plan. Investment was channelled into the defence and heavy industries, and away from education, housing and light industry (Lieberthal and Oksenberg 1988: 189). Yu Qiuli pledged rapid increases in petroleum production, which were met by the industry. Thus the MPI consolidated its position as the most powerful ministry in China, with its leaders counting among the top leaders of the entire country, due to its policy successes, which accorded with Mao Zedong Thought.

Ideology, rather than economic efficiency, continued to drive oil sector development through the 1960s. The Cultural Revolution in particular caused some disruption to China's oil surveying and exploration, refining, transportation and distribution arrangements (Kambara and Howe 2007: 24). However, despite the political turmoil oil production largely remained stable, mainly due to the oil field investments made prior to the onset of the Cultural Revolution (Lieberthal and Oksenberg 1988: 193). That being said, Daqing became a target for attacks by rebel workers; during the Cultural Revolution ministry-level cadres in industrial sectors were typically regarded as 'agents' and 'running-dogs' of capitalist-roaders (Hama 1980: 195). Red Guard activists such as Chen Boda took the Cultural Revolution right to the heart of Daqing claiming; "Daqing's red flag is actually black, its standard-bearers have falsified, its cadres are no good, its performances are deceptive" (Hama 1980: 195). Wells and equipment at Daqing were damaged, and the entire petroleum group came under attack from the Red Guard, with Yu Qiuli being the object of particular enmity from Jiang Qing (Mao's wife) (Lieberthal and Oksenberg 1988: 192). Though a complete worker takeover of Daqing never occurred, Iron Man Wang Jinxi was denounced and detained for a time – he was singled out as being the "number one political pickpocket of the whole country" (Hama 1980: 195). Model workers across the country were denounced as standard bearers of the black flag (imperialism), and the petroleum bureaucrats in Daqing were branded enemies of socialism. Zhou Enlai was incensed by these developments and moved to immediately reinstate Wang Jinxi and order to the field (Kambara and Howe 2007: 18). At the same time as these events unfolded, the focus of economic construction continued to be on the third front, which was pushed ahead rapidly between 1969 and 1971. During this time China had emerged from the 'narrow' Cultural Revolution and embarked upon a 'new leap forward' for third front development (Naughton 2007: 75). All efforts were geared towards industrial construction and the economy was effectively militarised – army officers were often placed in charge of production facilities (Naughton 2007: 76). As a result the country was on the brink of economic collapse by 1971. In addition, industrial growth outpaced agriculture by too wide a margin, placing great stress on food supply (Naughton 2007: 76).

The political and bureaucratic strength of the oil industry protected it from the worst effects of the Cultural Revolution (Kambara and Howe 2007: 25). Bureaucratically the industry was divided between the MOG and the MPI. The division of labour saw the MOG responsible for oil exploration, and the MPI in charge of production. With Mao's emphasis on maximising production output in the industrial sectors, power shifted in favour of the MPI, and it became increasingly responsible for both exploration and production (Kambara and Howe 2007: 25). Kambara and Howe (2007: 25) state that the growing clout of the MPI accounted for a concomitant slow down in the discovery of new proven reserves, since the MPI had less expertise and interest in oil prospecting. This led to an imbalance between production and exploration, and a consequent decline in the reserves to production ratio (Kambara and Howe 2007: 25). At the same time, the political prominence of the MPI, and the petroleum group more generally, was reflected in strong growth in the industry overall. Mao's continuing focus on defence and heavy industry meant that the influence of the oil industry on the country's economic planning during the socialist era remained high. However, the oil sector was fraught with inefficiencies and waste as the emphasis was on maximising output at any cost – there was no long-term plan for sustainable domestic oil industry development. Nor was actual oil production and consumption geared towards sustainable economic development, but rather to the establishment of a massive military-industrial complex. Some petroleum bureaucrats recognised this problem early on, but change would not be forthcoming because they continued to be hamstrung by the ideological and political commitments of Mao's approach to defence and industrial development.

The slow shift away from economic autarky in oil sector development

As early as 1962 the MPI recognised the need to import more advanced foreign technologies to enhance refining capacity and facilitate development of offshore oil reserves (Lieberthal and Oksenberg 1988: 181–183). The issue of oil refining was so pressing that China was willing to make an exception and tentatively turn to the outside world to alleviate the problem (Lieberthal and Oksenberg 1988: 195). Between 1963 and 1966 the Chinese signed contracts to purchase forty-six plants from ten western European countries and Japan (Lieberthal and Oksenberg 1988: 182). China's petroleum bureaucrats had conducted tours abroad to inspect western technologies and equipment in various exhibitions from 1963–1964, but on the whole the notion of systematically importing foreign technologies, especially for upstream exploration and production could not survive the Chinese political climate of the 1960s (Lieberthal and Oksenberg 1988: 194–195). Technological exchanges between China and Japan, and equipment purchases from France and Japan occurred sporadically between 1964 and 1970 (Lieberthal and Oksenberg 1988: 195). However, a genuine opening to the outside world did not occur and the oil sector in China remained essentially isolated. During the Cultural Revolution the strict policy of self-reliance was stressed to an even greater extent, making it impossible for the MPI to advocate purchase of foreign technologies and equipment (Lieberthal and Oksenberg 1988: 195). Lieberthal and Oksenberg (1988: 196–197) note that the guidelines for the annual plans and Fourth Five-Year Plan (1971–1975) "emphasized decentralisation and called for development projects that were 'small scale, indigenous, and labour intensive' rather than 'large scale, foreign, and capital intensive'".

The policy environment from 1970 onwards began to change, becoming more conducive to a turn outwards. Economic pressures necessitated the policy shift, but there were also some significant political events that facilitated China's opening to the outside world. Some of the staunchest supporters of economic autarky, defence budgets and militant opposition to the United States and the Soviet Union, namely Chen Boda and Lin Biao, were suddenly purged (Naughton 2007: 77).

Immediately thereafter a breakthrough in Sino-American relations occurred, marked by a visit to China by President Richard Nixon in 1972. Hence new opportunities for technology transfers arose as economic relations with the capitalist world were re-established (Naughton 2007: 77). Premier Zhou Enlai took the lead in crafting a more moderate course for China's economic policy and development. Purchases of foreign oil technologies and equipment increased from 1972, and in 1973 the State Council accepted a proposal drafted by Yu Qiuli entitled *A Request to Increase the Import of Equipment and Expand Economic Exchange*, which set forth a plan to import US$4.3 billion worth of equipment and whole plants (Lieberthal and Oksenberg 1988: 197). This was a significant step as it "bestowed legitimacy and coherence to the policy of foreign purchases" (Lieberthal and Oksenberg 1988: 197). This inevitably led to the next problem concerning how China intended to pay for these purchases. The solution was to start exporting oil to finance imports, which became an especially attractive alternative for all concerned in the wake of the 1973 Yom Kippur War and OPEC oil embargo (Lieberthal and Oksenberg 1988: 199). Despite the high prices China fetched for oil exports, balance of payments difficulties arose, which challenged China's desire to minimise its dependency on the outside world (Lieberthal and Oksenberg 1988: 199). Political struggles continued to plague the party leadership, precluding a fully coherent and comprehensive reorientation of Chinese policy. The impacts on the oil sector were minimal though, as economic expediency drove increased openings. However, it was not until Mao's death and the demise of the Gang of Four in 1976 that leaders in the oil sector were released from strict ideological constraints on industry development and oil policymaking.

The initial strategy for China's oil sector development was the product of the extant state capacities that the government had at its disposal, as well as deficiencies, that is, abundant labour and a dearth both of capital and modern energy technologies. Capacity was also built during the Maoist period in order to directly control and mobilise society toward the achievement of economic objectives, for instance, through commune organisation. Despite the fact that China had become self-sufficient in oil by 1963 as a result of the adoption of a quasi-military mass campaign model, the Chinese oil sector suffered from serious flaws that gave rise to economic problems over time. Already mentioned were the adverse impacts of an isolationist policy approach on the industry, in terms of its ability to access advanced technologies and equipment to enable more effective oil exploration and production. The focus on maximising output inadvertently encouraged a cavalier approach to oil exploitation that resulted in heavy inefficiency and waste, with no view to the longer-term sustainable development of the country's energy resources. The central government controlled all decisions over the entire oil commodity chain, from exploration through to distribution (Kong 2010: 8). Hence the government was also the only source of investment, and from the late 1970s this fiscal support began to dwindle (Kong 2010: 9). The mass campaigns themselves also had unintended adverse consequences. For instance, each campaign headquarters took on a variety of social functions, such as health care, education and day care, resulting in much higher total operations costs for the petroleum industry (Kong 2010: 9). As these problems began to surface oil output started to decline from the late 1970s, which was of great concern since the government had begun to use oil exports as a source of hard currency (Kong 2010: 9). In response, and in the context of a political climate more permissive to opening up to the outside world and adopting market-oriented approaches, the government embarked upon extensive reform of the petroleum industry.

With reference to economic development in China during the Maoist era, Oi (1995: 1134) claims "Ideology and the goals of state intervention, not an inherent failing in the policy instruments, undermined the capacity of the Maoist state to foster economic development." Hence it was excessive state intervention coupled with the emergence of a radicalised political environment that distorted incentives, resulting in suboptimal economic performance in the oil sector. Furthermore, Oi (1995: 1133) argues that the Maoist system left a legacy of useful policy instruments and political capacity at both the central and local levels, which formed the foundation for post-Mao economic development, particularly at the local level, "once local initiative and the proper incentives were introduced". Oi (1995: 1133) goes on to suggest, "Unlike late industrialising countries of Africa or Latin America that are often plagued by bureaucracies lacking experience or organisational capacity, the Maoist bureaucracy was an elaborate network that extended to all levels of society, down to the neighbourhood and work unit, and in international perspective it exhibited a high degree of discipline." That being said, the strong bureaucratic tradition in China certainly pre-dates Mao, characterising the Chinese political system for around 1,400 years (since the late sixth century there has been continuous unified rule of a single bureaucratically governed Chinese state) (Kroeber 2008: 30). Kroeber (2008: 31) claims that when the CCP began rebuilding its own institutions in the 1950s, and then again along more pragmatic lines in the 1980s, it was able "to appeal to this long bureaucratic tradition". He goes on to say "The history of China in the past 30 years is, to a significant degree, the history of the re-creation of a traditional bureaucratic ethos with an increasing degree of administrative effectiveness" (Kroeber 2008: 31).

While this study advances the argument that political elites in China can readily reorganise and reform institutions, it is important to note that historically accumulated institutional legacies also matter. With the shift from central planning to Reform Era marketisation we see the emergence of policies and institutional changes that are both historically accumulated and driven by party elites. In conclusion, and with reference to the onset of the Reform Era, which is dealt with in the next chapter, Oi (1995: 1147) claims, "Not only is the political strength of a regime on the eve of reform crucial to determining its capacity to structure economic change, but a regime must also ensure that it retains sufficient capacity to control the course of reform." Both institutions and incentives matter, and while continually building on existing capabilities and state structures, the central party-state has the authority and capacity to reshape and restructure them. Interestingly, institutional development from the Mao era through to the present day shows that China has continuously sought to balance a decentralising strategy, which produces greater flexibility, dynamism and initiatives at lower levels, and the need for a strong central state to coordinate and steer economic development. In other words, when redirected to economic modernisation during the Reform Era, Mao's approach of encouraging "decentralised initiative within the framework of centralised political authority" proved to be much more effective (Mirsky 2012). This becomes even more apparent during the Reform Era where the Chinese leadership attempts to balance the imperatives of marketisation and central party-state control. The next phase of oil sector development in China saw a much stronger emphasis on decentralisation, particularly of the market-oriented players (the NOCs) in order to overcome stagnating oil production and improve overall efficiency.

5
Decentralisation and Corporatisation of the Oil Sector (1978–2002)

Despite achieving initial success in oil exploration and production in the early 1960s, China's oil industry soon began to languish under the centrally planned economic system and radicalised political environment, both of which failed in the long run to provide the right incentives for new oil field discovery and production, and technological improvement. In 1978 Deng Xiaoping's 'second revolution', officially known as 'reform and opening' (*gaige kaifang*), marked the beginning of China's transition from a planned to market-oriented economy. This nationwide economic reform drive significantly restructured the oil sector. Specifically, oil production, the administrative apparatus and oil pricing were all decentralised to varying degrees. In addition, the ideological barrier to permitting foreign involvement in oil sector development was officially abandoned, hence China began to actively encourage FDI into the country and also sought to establish joint ventures with western oil companies. The decentralisation of oil production had the most profound impact on the oil sector as it saw the government delegate operational decision-making downwards to the newly created NOCs and largely withdraw from micromanaging oil production. The petroleum administration was decentralised with the abolition of the MPI in 1988 (its upstream assets were restructured into CNPC). This served to weaken central control and oversight of the oil sector, and at the same time strengthened the political clout of the NOCs, which assumed further decision-making authority for the industry in the absence of a central petroleum or energy ministry to coordinate policy. Furthermore, to improve the financial performance of state oil firms the government gradually relaxed oil price controls, especially for crude oil, with the aim of eventually integrating domestic and international oil prices. However, such integration should not be confused with genuine liberalisation, as it was an *ad hoc* government-controlled process, with the actual pricing mechanism remaining under the direct control of the country's top economic planners the entire time.

Overall these reforms delivered greater efficiency, profitability and production capacity to the oil sector. That being said, the oil sector reform process during this period was far from unproblematic, among other things creating a range of market distortions, which the central government then had to address through further reform. Hence this stage of oil sector development is best viewed within the context of market transition, which leads to a more nuanced understanding of how these market distortions are endemic to gradualist economic transitions and should not be portrayed as indicative of weak state capacity as such. The alternative of 'big bang' or shock therapy approaches to economic reform may have avoided such distortions, but would be much more likely to destabilise the entire party-state by creating a large proportion of 'losers', thus inviting fierce opposition to the reform process at an earlier stage during the first reform period. This chapter shows that the central party-state greatly improved oil state capacity through selective and carefully considered market-oriented reforms that led to creation of the NOCs. While there were many aspects of oil industry reform during this period that produced various market distortions and other suboptimal economic outcomes, these were dealt with reasonably effectively. It is important to note that across the two decades since the beginning of reform oil production and the profitability of the oil sector improved significantly.

Arguably state capacity was diminished in the area of policy decision-making and operational control of the oil sector. This was another consequence of decentralisation, but not one that was necessarily either unintended or unwelcome at that particular stage of oil industry development. Rather it can be viewed as a strategic move on the part of the centre to encourage the oil sector to build capacity in critical areas by withdrawing from its previous role of micromanager and exposing key parts of the industry to greater competition. Gradual decentralisation of oil production, administration and pricing was essential to replace the command-and-control system with a functioning market in the oil sector. Relinquishing operational autonomy to the NOCs was a necessary move in order to improve their performance, encourage the adoption of standards of global business practice, and generally prepare them to compete with international oil companies and other NOCs on the world stage. Furthermore, the key levers of central party-state control were retained, namely through the nomenklatura system, banks and powerful state planning institutions such as the SDPC (later renamed the NDRC), which remained responsible for oil pricing. During this period administrative capacity in the oil sector weakened, and while initially this was a deliberate decision taken by China's political elites, there was growing recognition by the end of the 1990s that the centre's ability to formulate coherent energy policies, effectively coordinate policies among multiple bureaucracies and guarantee their implementation had diminished, and as such became the locus of state capacity building efforts from 2003 onwards (the subject of the next chapter).

Foreign participation in oil sector development under reform and opening

As early as 1962 China's petroleum bureaucrats perceived the need to tap into foreign resources, namely technology and expertise to improve oil exploration and development (Lieberthal and Oksenberg 1988: 182). While limited *ad hoc* foreign oil equipment purchases and technology transfers occurred throughout the 1960s and 1970s, the political climate within China was not permissive to substantive and systematic engagement with the outside world. Minor moves that were conducted to aid oil industry development through foreign involvement were strongly condemned by radicals, and at times by Mao himself. Such condemnation was even directed towards China's decision, primarily endorsed by Premier Zhou Enlai, to export crude during the 1970s in order to earn hard currency (Kambara and Howe 2007: 26). Staunch opposition to foreign participation in China's oil sector served to slow its development, especially in the case of offshore oil exploration and production (Kambara and Howe: 27). By the late 1970s international engagement was the most essential requirement in order to stimulate oil exploration and production in general, and the exploitation of offshore oil in particular. Hence even more important than trade in crude and oil products was access to foreign technology and expertise.

The Sixth Five-Year Plan (1981–1985) sought active foreign participation in the oil industry, especially in offshore oil exploration (Arruda and Li 2003: 14). In 1978 a Japanese petroleum mission to China was conducted, largely for the purpose of explaining to the Chinese the legal implications of offshore exploration and the kinds of contracts that foreign oil companies were likely to offer, with a focus on articulating the benefits for China of engaging in production-sharing agreements (PSAs) (Kambara and Howe 2007: 28). Around this time a Chinese petroleum team embarked on a fact-finding mission to Europe and North America, visiting the British North Sea, Norway, Washington, New York, Houston, San Francisco, and returning home via Tokyo (Kambara and Howe 2007: 28). The main aim of this mission was to discuss issues surrounding offshore contracting. As a result of these information exchanges China undertook its first round of tendering for PSAs, signing contracts with Japan's Japan-China Oil Development Company and France's Elf Aquitaine for exploration and development in the Bohai Gulf and the Beibu Gulf in May 1980, and also with an American independent, Arco, for exploration of the Yinge Sea off the southern coast of Hainan Island. Initially the contracting party for the Chinese was the offshore sub-corporation of the Chinese Petroleum Corporation, which acted under the supervision of the MPI (Kambara and Howe 2007: 29).

Beijing had to construct an organisational and legal apparatus in a short period of time in order to host joint ventures with foreign oil companies and govern their involvement to China's advantage. Supply bases were developed to support offshore activities, and CNOOC was in 1982 as the designated legal entity to enter into joint ventures with foreign oil companies (Lieberthal and Oksenberg 1988: 123). The *Equity Joint Venture Law* was promulgated in 1979 to create a legal framework for foreign direct investment in China (Arruda and Li 2003: 14). This was followed by laws specifically governing the oil sector: the *PRC Exploitation of Offshore Oil Resources in Cooperation with Foreign Parties Regulation* in 1982 and the *PRC Exploitation of Onshore Oil Resources in Cooperation with Foreign Parties* in 1983, were enacted in combination with other regulations governing royalties and imports (Arruda and Li 2003: 14). The Ministry of Foreign Economic Relations and Trade (MOFERT) later renamed the Ministry of Foreign Trade and Economic Cooperation (MOFTEC), which was created in 1982, further regulated foreign economic activities in China. According to Arruda and Li (Arruda and Li (2003: 14)) these regulations "placed restrictions and conditions on foreign participation, and reflected the government's cautious stance on foreign trade and economic activities in China's energy sector". During the 1980s more than one hundred foreign companies participated in Chinese offshore oil ventures. Lieberthal and Oksenberg (1988: 265) claim that due to their "negotiating acumen" CNOOC and the MPI secured an impressive foreign investment of US$1.7 billion between 1979 and 1985 for oil exploration in the South China Sea.

Foreign participation also occurred in onshore oil development, and loans from the World Bank were used to upgrade the Daqing, Shengli and Zhongyuan oil fields (Lieberthal and Oksenberg 1988: 261). China was still capital starved, and realised its need for development assistance by pursuing several sources of finance including multilateral aid from the World Bank and the United Nations Development Programme (UNDP), and bilateral aid, primarily from Japan, for large capital-intensive projects (Lu and Todeva 2000: 7). From the mid-1980s poor results (particularly in offshore areas) and declining international oil prices discouraged foreign oil companies from participating further throughout the 1990s (Kambara and Howe 2007: 33). Declining world oil prices also resulted in a surge in Chinese oil exports in 1984–1985 as a way to increase earnings as prices fell, thus these exports continued to play a vital role in financing China's imports (Lieberthal and Oksenberg 1988: 263; Kambara and Howe 2007: 34). Despite these trends, substantive foreign participation and general interaction with the outside world proved to be a successful reorientation for China as it stimulated oil industry development, helped it to catch up to international standards of technology and petroleum business practice, and also earned China hard currency through oil exports to support the country's ever-increasing imports. Arguably China's need to encourage foreign participation in oil sector development provided the impetus for a major structural reorganisation of the industry, resulting in the creation of the NOCs, since foreign companies and investors required a single corporate entity with which to deal (Lu and Todeva 2000: 8). Importantly, these structural changes in China's oil sector map onto broader market-oriented transformations in the Chinese economy.

The creation of China's NOCs: CNOOC, Sinopec and CNPC

One of the main goals of China's reform of the state sector during the 1980s and 1990s was to separate government from enterprises, create new institutions to govern the market economy and transform the large state firms into modern corporations. This reform agenda extended to the oil industry, which was also a distinctive sector due to its strategic status. Because of this, the government's approach to the oil sector was even more cautious and controlled, as Kong (2010: 13) notes, "...before the central government withdrew from directly managing the petroleum economy, it wanted to ensure that its NOCs would be up for the task to guarantee the country's petroleum security as well as to preserve and enhance the value of state-owned assets." Hence the modernisation and corporatisation of China's oil sector was a heavily state-managed, incremental process. Since the beginning of the Reform Era, reform of the state apparatus responsible for oil production has consisted of two major restructurings (Kong 2010: 13). The first round of restructuring aimed to achieve the move away from central planning through the introduction of market characteristics into the oil industry, as the government sought to divest itself of oil production by creating SOEs (Kong 2006: 66). It entailed the abolition of the industry's line ministries, which were replaced by three NOCs; CNOOC, Sinopec and CNPC – a process that unfolded during the 1980s. The second stage of restructuring was geared towards the corporatisation and internationalisation of China's NOCs, and saw the vertical integration and partial privatisation of these companies' assets from the late 1990s to early 2000s, when China began to list parts of its largest SOEs on the international equity markets (Kong 2010: 13). This stage of restructuring was also part of a policy of 'grasp the large, release the small' (*zhua da fang xiao*), officially adopted in 1996, which aimed to downsize the state sector and develop the country's large SOEs, primarily those operating in strategic sectors, into internationally competitive conglomerates.

During the 1980s three NOCs were created to replace the industry's line ministries. CNOOC was established in 1982. It was placed under the authority of the MPI, and later under the State Council when the MPI was abolished. CNOOC was charged with offshore exploration and production, and offshore cooperation with foreign oil companies. Lieberthal and Oksenberg (1988: 123–126) claim CNOOC was an outgrowth of discussions and activities between Western oil companies and the MPI from 1978 to 1982 concerning China's offshore oil development. It was an independent entity specifically designed to "broker the country's relations with foreigners in the energy area", and was given exclusive jurisdiction over offshore oil activities in the areas assigned to it (Lieberthal and Oksenberg 1988: 123–124). In terms of bureaucratic rank, CNOOC was granted that of general bureau or vice-ministry, higher than a bureau but lower than a ministry. By 1985 this company had established offices in Houston, London, Tokyo and Hong Kong. Even though CNOOC was the designated entity that foreign firms dealt with on offshore oil issues and also possessed bureaucratic rank, the MPI and the State Council continued to be responsible for providing strategic direction and policymaking in this area (Lieberthal and Oksenberg 1988: 125). CNOOC deferred to these higher organs on policy issues, as opposed to technical and operational issues, which were within its purview.

In 1983 the State Council established the China Petrochemical Corporation, more commonly known as Sinopec Group, in order to consolidate the processing and distribution of petroleum products. The refining and petrochemical assets from the MPI and the chemical enterprises from the Ministry of Chemical Industry and synthetic fibre manufacturing from the Ministry of Textile Industry were brought together to create Sinopec. This company held ministry rank and operated directly under the State Council. Since the MPI had been stripped of its downstream functions, Sinopec also assumed responsibility for formulating policies for downstream petroleum activities. The remaining onshore exploration and production and administrative functions and governmental responsibilities, including social ancillary social services such as schools, medical facilities and transportation infrastructure, of the MPI were restructured into CNPC in 1988, which was also granted ministry rank. Furthermore, the State Council granted CNPC the right to oversee international cooperation in developing China's onshore oil resources.

It should be noted that there is another less prominent Chinese NOC, a more recent entrant into oil exploration and production field, known as the China National Chemicals Import and Export Corporation (Sinochem). This company was originally created in February 1950 as the sole state-owned oil trading enterprise (as part of the PRC's drive to establish a state monopoly on foreign trade and centralise trading authorities), and has followed a different corporate evolution to the NOCs that were launched in the 1980s. Up until the start of the Reform Era, the centralisation of trading authorities served political, as well as economic, functions: to promote China's imports and exports and the interests of its economic development, support China's "peaceful foreign policy" and "enhance the role of the proletarian dictatorship" in the struggle against international capitalism (Zhang 2003: 166). The national trading companies, such as Sinochem, set prices for exports, and were responsible for conducting import and export business according to a national foreign trade plan. Zhang (2003: 167) shows that from the 1950s to the 1970s Sinochem accounted for 12.8 per cent of the national trade total, thus establishing it as a significant player that enjoyed "a rather comfortable monopolising position in China's foreign trade".

Beginning in 1979, trade reform in China had a profound effect on trading companies such as Sinochem. The changing trade environment presented new challenges, as well as opportunities for the company, and the introduction of market competition in particular led this company to adopt a strategy of internationalising its business operations in 1987. The creation of Sinopec and CNPC soon began to threaten Sinochem's monopoly over crude trading between China and the outside world. The company was also under pressure to expand its business scope from simply being an oil trading company to a fully integrated oil company. This involved incorporating both upstream and downstream capabilities. Since domestic upstream and downstream assets, both onshore and offshore, were already controlled by Sinopec, CNPC and CNOOC, Sinochem had to look abroad in order to acquire crude and refining assets. Sinochem was excluded from the 1998 restructuring of the NOCs, which further impeded its attempt to transform itself into an integrated oil company domestically. Since 2002 Sinochem has been authorised to invest abroad in oil exploration and production (Arruda and Li 2004 (February): 26). Thus it began to acquire significant upstream assets on the international market, but acquiring downstream assets has proven to be more difficult. That being said, upstream oil activities remain a smaller component of Sinochem's overall operations, which focus more on the import and export of petroleum and petrochemical products, oil refining, oil transportation and storage and the production of petrochemical products through its chemical business (Arruda and Li 2004 (February): 26).

CNOOC has always been the most commercially-oriented NOC, particularly since it was set-up to deal with foreign firms and hence was not saddled with additional bureaucratic and social functions, as was the case with CNPC and Sinopec (Chen 2007: 49). As a result CNOOC has consistently occupied a far better financial position than the other NOCs. Kong (2010: 14) claims that CNPC and Sinopec should not be considered standard SOEs, since they were all placed under the direct supervision of the State Council and tasked to perform the administrative functions of their predecessor line ministries. In addition, they also inherited bureaucratic rank, and were not simply market participants, but also functioned as market regulators (Kong 2010: 14). In the absence of an energy ministry to provide strategic direction and coordinate policy the NOCs were left to their own devices, and CNPC in particular was tasked with providing policy direction for the industry. The nominal separation of the central government and SOEs was further enhanced in 1992 when the *Regulations for Transforming the Operating Mechanism of the state-owned Enterprises* were promulgated (Zhang 2003: 97). These decentralising regulations were intended to grant SOEs operational autonomy by severing the link between the state firms and the bureaucracies they operate under. Under this regulation decision-making power was formally ceded to state firms. Management autonomy was extended to production, pricing, distribution, purchasing, import and export, and investment and finance, and was protected by law (Zhang 2003: 97).

Furthermore, in order to protect those state firms under the direct control of the central government from bureaucratic intervention, the enterprises were granted 'separate planning status' (Zhang 2003: 97). Despite the formal separation of government and enterprise on paper, the boundary in practice was of course much less clear. While the NOCs possess operational autonomy, their strategic direction was still subject to party-state influence and intervention. There is of course a range of political and organisational mechanisms used by the Chinese government to solicit policy compliance, which have already been mentioned and include the nomenklatura system of party personnel appointments (administered by the COD), central planning agencies and the CCP groups and committees established within state firms. Top executives and senior management of the newly created NOCs were former petroleum bureaucrats and party members of significance. For example, the former minister of the MPI, Wang Tao, became the head of CNPC. Kong (2010: 22) argues that this allowed the NOCs to influence the policymaking process, hence enhancing their political clout: "these informal connections offer oil companies direct access to the country's party leadership, sometimes bypassing the routine bureaucratic procedures and enabling them to secure critical endorsements for projects promoted by oil companies." This further indicates the lack of genuine separation between business and government in China's oil industry.

The functional specialisation of each NOC, that is, CNOOC in offshore oil development, Sinopec in oil refining and CNPC in onshore upstream exploration and production, was a legacy of ministerial compartmentalisation, which had been transferred to commercial operations through the creation of separate upstream and downstream conglomerates (Zhang 2003: 183). This compartmentalisation posed a challenge for the oil companies in terms of developing their international competitiveness, and also made them particularly vulnerable to oil price volatility. There was a perceived need to weaken the upstream/downstream monopolies, by establishing an oligopolistic market structure where several market players could compete. The market structure that was established in 1998 was specifically designed to promote limited and managed competition among the NOCs, and also encourage them to seek profits in both upstream and downstream activities (Kambara and Howe 2007: 117). This partly contributed to the further transformation of the NOCs into vertically integrated enterprises. In 1998 CNPC and Sinopec were directed to swap assets along geographical lines so that each company gained both upstream and downstream portfolios, with rough geographical monopolies. This move provided Sinopec with an upstream portfolio concentrated in China's south and CNPC with refineries and a distribution network in the north. Each company remained dominant in their original segment of the oil market – CNPC in onshore upstream exploration and production, Sinopec in oil refining and CNOOC in offshore oil, so as to avoid creating too much head-to-head competition. The market position of each affects their financial performances (Houser 2008: 146). For instance, since upstream oil prices were liberalised in the 1980s and closely follow global price movements, CNPC has benefitted immensely from the rise in world oil prices since 2003. Downstream oil prices, on the other hand, remain tightly controlled, and in some years, when domestic pump prices were below that of international crude oil prices, oil refining became a loss-making exercise for Sinopec. In order to further enhance the efficiency and competitiveness of the NOCs, the government deliberately introduced limited competition among the three oil companies through their vertical integration, and by establishing an oligopolistic market structure within which they operated.

Despite these reforms, the NOCs still had "bloated organisations, redundant personnel, heavy burdens of debt, and low-quality assets" (Wu 2005: 97). In order to transform them into modern competitive enterprises, the better performing core assets were carved out of the original or holding companies and restructured into joint stock limited companies for IPO financing (Wu 2005: 97). In other words, it involved separating the core from the non-core business, and corporatising the resulting entity containing the core assets. The historical burdens of non-core assets, non-performing financial claims, redundant personnel and other employee-support functions, and controversial projects such as CNPC's holdings in Sudan would be left to the parent company (Wu 2005: 97). For instance, at CNPC the number of employees in the core company that were floated through PetroChina was 480,000, compared with 1.6 million in the holding company. Employment is yet another social responsibility undertaken by China's large SOEs, as the government is concerned about the potential for social unrest caused by unemployed workers (Pearson 2007: 725). Hence this final stage of the restructuring saw the NOCs transfer their performing assets to subsidiary companies, which were subsequently listed on international stock exchanges. PetroChina, the core assets of CNPC Group, was listed on the New York and Hong Kong stock exchanges in April 2000; Sinopec Ltd, the core assets of Sinopec Group, was listed on the New York, London and Hong Kong stock exchanges in October 2000; and CNOOC Ltd (incorporated in Hong Kong, unlike PetroChina and Sinopec which were incorporated in the PRC), the core assets of CNOOC, was listed on the New York and Hong Kong stock exchanges in February 2001. The Chinese government remains the majority shareholder in all cases. Since 2004 CNPC and Sinopec acquired offshore exploration and development rights, and CNOOC has been given land-based exploration and development rights (Guo 2007: 3).

These listings gave the NOCs the capacity to raise funds through international capital markets, to be invested in further exploration, production and refining projects. In addition, it was intended to promote corporate governance among Chinese companies by providing "the managers with clearer incentives to focus on profitability" (Andrews-Speed 2004a: 179). The cultivation of 'national champions' through overseas listings also advances Beijing's desire to establish Chinese brand names that can compete in global markets (Dickson 2011: 39; Ewing 2005: 329). Soon after China's NOCs became listed, Beijing promulgated the 'Go Out' (*zou chuqu*) policy, encouraging the NOCs to acquire equity stakes in oil and gas production abroad. This policy was largely intended to improve the country's energy and natural resource security. Hence, it is clear that throughout thirty years of reform China's oil industry has been consistently regarded as far too important to be left to market forces or for the Chinese government to establish a genuine arm's-length relationship, let alone to consider extensive privatisation (Pearson 2007: 724). In other words, the development of this industry has occurred through top-down reform, rather than exposure to market forces. While they were products of this state-managed corporatisation process, CNPC and Sinopec assumed operational autonomy and the ability to influence policy direction, a development further reinforced by the transferral of former petroleum bureaucrats to top management positions.

The oil price reform conundrum

Oil price reform has been one of the most difficult and controversial elements of China's oil governance regime. This is because the price of crude and refined oil products impacts not only upstream and downstream profitability in the oil sector, but almost every other sector of the Chinese economy, and has the ability to fuel inflation. Hence Beijing has pursued a very cautious and incremental approach to oil price decentralisation. Again, the wider context of market transition is necessary to understand the oil price regime in China. In response to the various problems that have arisen from inevitable market distortions that accompany state set pricing, especially in the transition from a planned to a market economy, Beijing's energy planners have swung back and forth from exerting state control to variable degrees of market deregulation of oil prices since the beginning of the Reform Era (Chen 2006: 169). While oil price levels have been the subject of reform and continual adjustment (more recently in line with international prices), the actual pricing mechanism remains strictly controlled by the central government. Complete deregulation of oil pricing has not been on the agenda because Beijing wishes to maintain oil prices within a tight band that is sufficient to support the domestic oil industry, while at the same time low enough as to avoid any threat to economic growth and social stability (Downs 2004a: 180–181).

From 1949 to 1981 the oil pricing regime was simple and straightforward: oil was sold at a single, state-controlled price, regardless of oil quality (Chen 2006: 157). A two-tiered price system was introduced in 1981 in order to address capital constraints resulting from state-controlled low prices, which were limiting the capacity of China's NOCs to undertake oil exploration and new oil field development with the overall effect of reducing oil production (Kong 2010: 10). The two-tiered price system entailed a state directed national oil allocation plan, and an additional market for surplus oil production, which followed international pricing parity. This two-tiered price system was a key instrument of the dual-track system (*shuanggui zhi*), which refers to the coexistence of two coordination mechanisms (plan and market) operating in the state sector and industrial economy at large, ensuring a more stable transition from central planning to a market economy (Naughton 2007: 92). Laffont and Senik-Leygonie (1997: 19) note that in addition to eliminating the most serious shortages, the dual pricing system also enabled state firms to "adapt progressively to the laws of the market". In applying this system to the oil industry, the government adopted a contract responsibility model. Under this model a centrally mandated annual production target of 100 million tonnes was set, and 94.5 per cent of actual oil output was appropriated by the central government (Kong 2010: 10). Surplus oil production could then be sold abroad through Sinochem, at international prices, or domestically at market prices (Chen 2006: 157). The foreign exchange earned from the sale of surplus oil output was then used by the MPI to fund domestic oil exploration and production, and purchase foreign technology and equipment to aid oil sector development (Kong 2010: 11). By 1982 this system had begun to reverse the trend in declining oil production (Kong 2010: 11).

In terms of the pricing regime for the centrally mandated upstream oil production and allocation, there were two state-set pricing systems in operation. First, there was an in-quota low price for subsidised oil (300–500 yuan/ton, depending on the quality of oil, US$4.25–8.25/bbl equivalent at the exchange rate of 8.3 yuan per dollar), which served the military, agriculture and some large oil-consuming state firms (Wang 1999: 56). Second, there was an in-quota high price (about 700 yuan/ton in 1993, US$16.82/bbl equivalent, and 964 yuan/ton in 1997, US$15.91/bbl equivalent) (Wang 1999: 56). Kong (2010: 11) claims that in comparison to international oil prices, the prices of crude oil in China were still very low – in 1987 accounting for only 43 per cent of the international price. Goldstein (1992: 15) calculates that from 1972–1988 subsidised oil was sold at Rmb 100 per tonne, equivalent to US$6.60 per barrel across that period. By 1992, it had been increased to Rmb 200 per tonne, equivalent to US$5 per barrel (Goldstein 1992: 52). Thus, for more than two decades China's oil price had been decreasing (Goldstein 1992: 52). At the same time, oil development costs in China were increasing: labour costs were rising due to the additional social functions the NOCs were expected to fulfil, and production costs were increasing as the country's oil fields were aging (Kong 2010: 11). Oil development costs consequently increased by 70 per cent between 1986 and 1990 (Kong 2010: 11). Due to its overwhelming specialisation in upstream oil exploration and production CNPC was particularly cash-starved and was losing as much as Rmb 6 billion per year (Goldstein 1992: 53).

The upstream sector also suffered as a result of the pricing distortion that favoured downstream oil products (Kong 2010: 11). Crude oil price reform was outpaced by price decontrol in refined oil products. Since 1982 there have been two price levels for oil products: in-quota low prices and market prices for products once the centrally mandated quota had been met (Chen 2006: 157). The proportion of market-refined products increased significantly from less than 10 per cent in 1983 to about 65 per cent in 1993. By 1993 two-thirds of China's refined oil products were being sold at or higher than market prices. The consumers of refined oil products, with the exception of those on the receiving end of state-controlled low prices, actually paid high prices by international standards. For instance, petrol sold for around US$45 per barrel, which was higher than US prices, though lower than those of Japan and Europe (Goldstein 1992: 53). Furthermore government subsidies were only left for a few consumer categories, such as diesel fuel for farmers, fuel oil for certain state firms and refined products for military use (Wang 1999: 630). As a result refineries expanded in order to capture the benefits created by the price ratio imbalance, and at the same time the upstream sector shrank due to lack of investment (Kong 2010: 11). The country's largest oil refiner, Sinopec, was flush with cash since it was "the main beneficiary of a system that ensured artificially low crude prices and relatively high prices for refined products" (Goldstein 1992: 53). Patrick Cheung, an oil industry consultant at McKinsey & Company in Hong Kong, says "China historically chose to keep refining margins very high by international standards to ensure sufficient capital flow into building refining capacity" (Goldstein 1992: 53). However, the unintended consequence of this was the stagnation of crude oil production.

The central government took steps to get the oil industry out of the red by initiating several price increases from 1990–1992 (Chen 2006: 157–158). Despite the dire situation for the oil industry, Beijing was more concerned about the impact the oil price regime would have on inflation. Hence the approach to oil price reform was again state-controlled and gradual. By the early 1990s the efficacy of the entire dual track system was being questioned because it resulted in the inefficient allocation of resources and had outlived its usefulness in assisting state firms to adapt to market conditions. Deng Xiaoping's Southern Tour in 1992 marked the beginning of accelerated and more extensive price decentralisation – price ceilings for many goods were removed, and the State Price Administration reduced the number of production goods it regulated from 737 to 89 (Garbaccio 1995: 3). Crude oil prices were among those that were deregulated to a great extent. By 1994 domestic crude prices were equivalent to 77 per cent of international prices (Chen 2006: 157). In 1994 the central government abolished the contract responsibility model and integrated the in-quota and above plan oil prices (Kong 2010: 12). The policy shift to market deregulation for crude and refined oil prices in the early 1990s was prompted by both the national marketisation drive promoted by Deng, and problems specific to the oil sector, namely oil production stagnation, inefficiency and waste in the allocation of energy resources, a dearth of capital for investment in upstream oil sector development and CNPC's massive financial losses.

This 'oil shock treatment' proved to be unworkable in an environment where the Chinese leadership was attempting to balance the demands of reform, stability, development and growth, and many adverse consequences for the oil industry and the economy as a whole ensued (Chen 2006: 159). Sinopec was plunged deep into debt, with its earnings in 1993 slashed by 50 per cent (Goldstein 1993: 50). After the State Council decided to abolish low plan-price oil, which averaged around Rmb 265 per tonne, Sinopec's price for 35–40 per cent of its crude oil supplies skyrocketed by 134 per cent (Goldstein 1993: 50). While retail prices for oil products also increased, the proportion could not match the rise in crude oil price. Furthermore, as a legacy of both the Maoist era of self-sufficiency where every province was expected to have its own refining capacity, and further expansion of this refining capacity when crude oil was set at a low price, Sinopec was saddled with too many refineries, most of them of insufficient size to be economically viable (Goldstein 1993: 50). The increase in crude and refined oil prices also contributed to high rates of inflation, jumping to a 'perilous' 26 per cent in February 1994 from 23 per cent in January, much higher than the desired rate of less than 5 per cent (Chen 2006: 159). High inflation not only threatened social unrest, but also fed into higher oil production costs (Chen 2006: 159).

These developments presented Beijing with a stark policy dilemma, as Chen states, "on one hand, China needed to curb the rising prices so as not to compromise economic growth and social stability, but on the other hand, if China kept the refined oil price lower than the world price, then the petroleum industry could hardly recover from economic losses" (Chen 2006: 160). The solution was a return to the command approach in 1994; China reasserted central control over the oil sector "by fixing the price of crude and petroleum products and channelling virtually all sales through state agencies" (Wang 1999: 627). Specifically, the SETC and SDPC assumed the role of setting both crude and refined oil prices. This return to a state administrative, rather than the market solution was driven by the need to ensure social stability, which was a clear priority for the newly established Jiang Zemin administration (Chen 2006: 162). Chen (2006: 161) explains "The SPC and SETC took charge of oil production and distribution; the major channels for refined oil distribution were provided by Sinopec with the assistance of CNPC and local refineries." However, the measures taken by the central government did little to stem inflation, leading experts to conclude "The oil price controls appears to have only one obvious beneficiary: the state oil monopoly" (Chen 2006: 162).

From 1998 to the present the central government has focused on gradually bringing domestic oil pricing closer to international pricing. This decision was taken as part of the broader overhaul of the oil industry, which included a substantial restructuring of the NOCs. Chen states, "Since China endeavoured to join the WTO and had to import oil from overseas, the Chinese energy market was set to be connected with the world one" (Chen 2006: 163). This round of oil price reform was largely intended to further expose the NOCs to market forces in an effort to make them more competitive on the international stage. A new mechanism was introduced in June 1998 to link domestic crude and refined oil products to world oil prices when the State Council promulgated the *Plan of Crude Oil and Refined Product Price Reform*. Under this mechanism crude oil pricing followed a formula more 'reasonable and open' than the two-tiered pricing system used up until May 1998, albeit set in a government-controlled fashion. In this new system the SDPC formulates the price of crude oil according to a marker price plus a premium (*Oil & Gas Journal* 1998: 46). The marker price is based on the freight on board (FOB) price of international crude of a comparable grade, plus a tariff set at 16 yuan/metric ton (*Oil & Gas Journal* 1998: 46). China's domestic crudes were classified into four grades: Light Crude, Intermediate Crude I, Intermediate Crude II and Heavy Crude (*Oil & Gas Journal* 1998: 47). CNPC and Sinopec monitor the Singapore market and submit to the SDPC the daily strike prices of different crudes for the previous month (*Oil & Gas Journal* 1998: 47). The SDPC then produces a marker price for the month by averaging the month's strike price plus the premium (*Oil & Gas Journal* 1998: 47). The marker price then becomes effective throughout the next month. The premium is determined by the crude transportation cost and price differentials among the various crude grades (*Oil & Gas Journal* 1998: 47). CNPC and Sinopec are permitted to move the marker prices 5 per cent up or down at the petrol station (*Oil & Gas Journal* 1998: 48).

Despite this commitment to integrating domestic and international oil prices, Beijing remains resolute in preventing oil prices from either climbing too high or falling too low, which would either drive inflationary pressures or erode the profitability of the domestic oil industry. The SDPC also indicated that domestic crude oil should be sold at a slightly lower price for the purpose of encouraging domestic refineries to purchase Chinese crude (*Oil & Gas Journal* 1998: 47). In 2001 the SDPC decided to link domestic oil prices to the three global trading hubs of Rotterdam, Singapore and New York, using their weighted averages to formulate the marker price, and permitting CNPC and Sinopec to set retail prices within a range of 8 per cent of the published guidance price. This was a solution to a problem that had arisen with the previous pricing mechanism being excessively transparent, and thus inviting over-speculation in the domestic oil market. However, this new oil pricing system has not been problem-free either. For example, China lags behind fluctuating international oil prices as a consequence of this mechanism (Kong 2010: 13). As a result Chinese oil refineries have an incentive to export their oil products during periods of high international oil prices, and this could result in domestic oil shortages (Kong 2010: 13).

The history of China's oil price reform since the beginning of the Reform Era shows that oil serves not only economic purposes but also political ones. Beijing's political objectives are concerned primarily with ensuring the 'smooth running' of the economy so as to prevent social instability. Suppressing oil prices offers an "industry-wide subsidy" and serves to manage inflation, which is a priority for the Chinese leadership (Ma 2013). When measured against purely economic indicators, China's oil price regime may appear to introduce problematic market distortions, and place a heavy financial burden on either state coffers or the NOCs during times when world oil prices are high. However, China possesses the fiscal capacity, especially since the late 1990s and 2000s, to subsidise certain oil consumers and compensate oil refiners when the need arises, and this is perceived within Chinese policymaking circles to constitute a preferable approach for the time being. Since the PRC's inception, Beijing has shown a consistent policy preference for controlling the price of oil and has been prepared to endure and deal with the various distortions and perverse incentives to which a state-managed oil price regime gives rise. The alternative of full oil price liberalisation and elimination of oil subsidies would introduce greater efficiency into the domestic oil market, but would also compromise economic growth and pose an unacceptable level of risk to social stability (and hence to the ultimate survival of the CCP, which derives a large part of its legitimacy from maintaining order) by driving up inflation, especially when international oil prices are high. Therefore from a state capacity perspective China has the means to effectively pursue the current oil price regime, which appears appropriate given the range of political objectives to which the party leadership currently gives priority, and the central role of oil in meeting China's growth objectives. Furthermore, control of oil pricing is fundamental to Beijing's ability to exert control and set policy for the oil industry as a whole in China.

Bureaucratic reform and decentralisation

Under the centrally planned system the Chinese oil industry's pursuit of quick results in oil production, rather than engaging in longer-term planning, had resulted in declining oil output by the early 1980s. The focus on rapid oil extraction produced a widening imbalance between resources allocated to maximising output in proven oil fields and those providing for prospecting of new oil fields (Kambara and Howe 2007: 29). This was also partly due to the expansion of the MPI's power and influence at the expense of that of the MOG, which was responsible for oil prospecting (Kambara and Howe 2007: 29). In order to reverse this trend the government embarked upon administrative reform, and attempted to install an "improved structure of material incentives" in place of extant political incentives (Kambara and Howe 2007: 30). In the era of central planning "regulatory bodies were formed with dual responsibility for the production of energy and its regulation," and the oil industry essentially functioned as an appendage of the government (Arruda and Li 2003, 2004: 19). Importantly, production and regulatory functions were separated during the reform period, with the administrative infrastructure being further rationalised throughout the 1980s and 1990s. Arruda and Li (2003, 2004: 19) claim that the earliest recognition of the need to split administration from commercial activities in China occurred in the oil industry. China's oil sector administration was decentralised as a part of broader reform initiatives that occurred at the national level. Hence the impetus for such reforms did not solely emanate from the oil sector itself, rather it was part of a nationwide drive to marketise the economy and shift away from command-and-control regulation. These reforms reduced the Chinese government's centralised control over the oil industry. In 1988 the MPI was abolished, and with its downstream oil functions already transferred to Sinopec, the remaining upstream oil functions and administration were restructured into CNPC. Both Sinopec and CNPC possessed ministry rank and regulatory capabilities, while CNOOC was placed under the direct supervision of the State Council once the MPI was abolished.

The Ministry of Energy (MOE) was established in 1988 in order to oversee the oil sector in addition to other energy sectors. The MOE consolidated the administrative functions of the MPI, the Ministry of Coal Industry, the Ministry of Nuclear Industry and the power sector of the Ministry of Water Resources and Electric Power. According to Downs (2006: 17) these ministries opposed the merger, with officials from the former Ministry of Coal Industry petitioning to have their ministry reconstituted. The MOE was really only active and effective in the electricity sector, as the other energy sectors refused to coordinate planning and investment activities (Downs 2006: 17). In addition, the authority of the MOE overlapped with that of the SDPC and the state energy companies. Premier Zhu Rongji made the decision in 1993 to abolish the MOE, partly due to these deficiencies and overlapping functions, and also as a part of a wave of further administrative retrenchment undertaken at that time (Downs 2006: 17). In the wake of the MOE's abolition ministries were reestablished for the coal industry and electric power however, none were reestablished for the oil industry, which now lacked both a ministry with exclusive oversight of the oil sector and a more general energy ministry. Instead the oil industry was primarily governed by two macro-regulatory agencies – the SDPC and SETC. The SDPC was the most powerful actor responsible for energy policies and remained in charge of investment approval and pricing, whilst the SETC, though nominally equal with the SDPC, played a more marginal role in the oil sector (Andrews-Speed 2004a: 175). The NOCs were able to assume policymaking functions due to their ministry-level status. This bureaucratic rank meant they were directly accountable to the State Council, rather than to other ministries and possessed a dual role as government organ and commercial enterprise (Andrews-Speed 2004a: 176). CNPC took the lead for oil exploration and production, and served as the main policy advisor to the State Council on petroleum issues.

In 1998 the State Council created the State Petroleum and Chemical Industry Bureau (SPCIB) under the SETC to reassume some of the administrative functions that had been delegated to the NOCs. At the same time these companies were being prepared for floatation, which required the removal of non-commercial operations, and their structural reorganisation into regional vertically integrated companies. Andrews-Speed (2004a: 178) claims that the integration of domestic and international oil prices in 1998 also marked "a significant step forward to provide incentives to the Chinese state oil companies to improve their technical and financial performance." After the 1998 reforms the oil industry was regulated by the Price Administrative Department of the SDPC, the Transport and Energy Department of the

SDPC, the SETC and the Ministry of Land and Natural Resources (Andrews-Speed 2004a: 178). In addition to being responsible for investment approval and pricing, the SDPC also secured the role of long-term policy formulation for the energy sector after 1998 (Andrews-Speed 2004a: 176). The SPCIB was granted only bureau-level status, and hence did not possess sufficient authority to direct the NOCs. In terms of the SPCIB's responsibilities, Andrews-Speed (2004a: 176) states that they were "never clear to the outside observer, though they appeared to include policy and regulatory tasks, as well as aspects of enterprise management." Kong (2010: 16) claims that as a result of this ambiguity the SPCIB would frequently defer to the NOCs over policy problems. The largely ineffectual SPCIB was abolished in 2001 in a move to further deregulate the oil industry, and replaced with an industry association known as China Petroleum and Chemical Industry Association (CPCIA). Therefore specific industry oversight of the oil sector became progressively weaker and more fragmented, a consequence of a deliberate move by the central party-state to marketise various sectors of the Chinese economy. From 2003 the centre began to reverse this trend in a gradual series of attempts to recentralise political authority and facilitate better coordination among bureaucracies in the oil sector (the subject of the next chapter). However, from 1993 up until 2010, when the National Energy Commission (NEC) was established, there were no ministry-level organs in operation to oversee the oil sector, and the institutional set-up was indeed characterised by fragmented and often unclear lines of authority.

China's oil industry in the first Reform Era

When evaluating state capacity in China's oil sector from the start of the Reform Era up until the end of 2002 (when energy security suddenly rose up the Chinese leadership's political agenda), the context of market transition must be acknowledged. Oil sector reform during this time occurred within the early stages of a complicated transition from a planned to a market-oriented economy, and as such was nested within a particular policymaking framework governing this transition and long-term plans for economic development. Since it is a strategic sector of the Chinese economy the oil industry was never intended to 'grow out of the plan' in a manner similar to the country's nascent non-state sector (Naughton 1996). Rather it was expected to adopt market characteristics in order to improve performance and efficiency, while at the same time remaining under the direct control of the central party-state. China pursued gradualism in its approach to economic reform, rather than the big bang or shock therapy approach advocated by neoliberal economists and Washington-based financial institutions. The main advantages of gradualism include its ability to create a constituency in favour of reform, since it cushions the harsh blow of transition and avoids creating too many losers at once. The use of the dual track system in particular is often cited as a prime example of 'reform without losers' (Pei 2006b: 26). Accordingly partial reform precludes the emergence of fierce opposition, and instead creates a sustainable momentum for further reform. Hence 'reform without losers' was a key concern during the first Reform Era, but came to an end in the mid-1990s when the Chinese leadership embarked on reform of the state sector aiming to return it to profitability through downsizing and streamlining, which resulted in extensive layoffs (Naughton 2008b: 121–122). Gradualism also affords policymakers greater flexibility and learning capacity, as captured by Deng's axiom 'crossing the river by feeling for the stones'. In explaining the logic that underpins gradualist reform, Pei (2006b: 23) states, "Gradualism allows decision makers to target certain sectors for breakthrough reforms and acquire valuable knowledge for applying reform to other sectors. Most importantly, gradualism allows reformers to make – and correct – policy errors and avoid costly mistakes that can fatally undermine the support for reform."

The flipside of gradualism is that it often results in a lack of complementarity (where some reform measures cannot be fully effective without other accompanying reforms), which can distort markets and provides opportunities for rent-seeking behaviour by party officials (Pei 2006b: 22). Pei (2006b: 28) notes that Fan Gang, Professor of Economics at the Chinese Academy of Social Sciences (CASS) and a proponent of gradualism, "admits that gradualism carries huge costs, especially in terms of efficiency losses, continuing price distortions (due to the controls imposed by the government on key inputs), soft budget constraints, and monopoly." Some of these costs were certainly evident in the first two decades of oil sector reform, however, when evaluated against successful capacity-building efforts in other areas of the oil industry, and taking into account the centre's ability to overcome these problems through further reform and learning from past mistakes, then the scorecard for state capacity in the oil sector appears to be much more positive. Hence oil sector reforms during this period were, on balance, largely successful. Before 1978 the central government controlled all aspects of the petroleum commodity chain, and central bureaucracies were responsible for oil production, with the government constituting the sole source of investment. The central party-state's commitment to industrial modernisation led to the loosening of various political and economic constraints that had previously hindered oil sector development. Within the space of several years the line ministries were abolished and the NOCs were created. These companies were able to navigate a steep learning curve and were progressively granted operational autonomy as they underwent corporatisation and adapted quickly to market conditions. Beijing's main aim towards the oil sector during this time was to improve the financial and technical performance of the NOCs in order "to increase net economic benefit to the nation" and enhance the country's energy security. Hence much effort was geared towards creating better incentives and a system of regulation in order to provide a basis for the commercial transformation of China's NOCs (Andrews-Speed 2004a: 171). The second round of restructuring in 1998 was driven by the perceived need to internationalise state firms. These reforms further improved the performance and efficiency of the NOCs and encouraged them to compete within China, albeit in a very limited manner, boosting their profits and international competitiveness.

The separation of government and commercial activities in the oil industry, coupled with administrative decentralisation and retrenchment, was a deliberate strategy adopted by Beijing with the aim of improving the overall production, profitability and efficiency of the oil sector, mainly through encouraging the commercial development, and subsequent internationalisation, of the NOCs. This reform drive is neatly summarised by Rosen and Houser (2007: 18): "By the end of 2002, all energy production and delivery in the country was being carried out by companies rather than bureaucrats, and these firms were making investment decisions based *largely* on market, rather than political considerations." The centre's delegation of operational and technical autonomy downwards to the NOCs was essential to achieve these goals. However, it should be noted that although the separation of government and the commercial management of state firms was achieved on paper, in practice the separation was not as clear and pronounced due to the extant levers of control the central party-state wields throughout the strategic sectors of the economy. Importantly, the state has always maintained its strategic control of the oil industry. That being said, the centre's ability to formulate coherent policy and coordinate implementation within the oil sector diminished as a result of administrative decentralisation. Up until the early 2000s this weakened capacity was not particularly problematic since energy security occupied a position much lower down on the Chinese leadership's policy agenda. While the party-state retained strategic control of the oil industry, the NOCs were able to influence policy direction. This is because the NOCs were run by former petroleum bureaucrats who possessed direct links to the party leadership, thus enabling them to bypass any intermediary bureaucracies. The NOCs themselves were repositories of specialised industry knowledge, critical to effective governance of the oil sector, and as already stated, the Chinese leadership did not have its mind on energy security at this time.

This has led some scholars, such as Kong (2010: 23), to suggest that the decentralisation of oil production, prices and administration gave rise to a 'co-governance structure', whereby both the NOCs and the central government are deemed to shape oil policies. However, the power relationship between the central government and the NOCs is certainly not one based on equality, as might be implied in Kong's conceptualisation. While the NOCs play an advisory role in terms of policymaking, the government is the sole authority that provides strategic direction and is in charge of decision-making in the oil sector. Various levers of control ensure the party-state's authority over the oil industry in general, and also towards the NOCs. These include control of oil pricing, the investment approval system, state ownership of the NOCs, the nomenklatura system and Party groups and committees that are installed in state firms. Taken together these powerful instruments ensure the Chinese state's control of the oil sector in China. They are also sufficient to control the NOCs, even where state interests and the interests of the NOCs diverge – the imperatives of the party-state consistently override corporate considerations (this will be discussed further in Chapter 7).

Overall then the decentralisation push during the first Reform Era, while necessary to achieve a functioning market where none existed before, fragmented the oil sector and weakened state capacity. Decision-making authority in particular became fragmented among multiple bureaucracies, without the establishment of a clear hierarchy. This was all part of carefully considered decisions taken by China's political elites to introduce more market-based incentives into the oil industry by decentralising oil production, administration, and, to some extent, pricing; hence it is best understood as an elite-driven process where the central party-state relinquished its previous role in micromanaging oil activities. In the absence of a clear bureaucratic hierarchy in the oil sector or major input from the party leadership, the NOCs (particularly CNPC) assumed responsibility for the strategic direction and development of the oil industry. Furthermore the oil industry expertise that had previously been "housed" within the industrial ministries was largely transferred to the NOCs, creating a "policy vacuum in Beijing" and reducing policymaking capacity (Rosen and Houser 2007: 18). As a result a narrow emphasis on expanding oil supply became the primary focus of China's oil policy approach, because it was driven by the more commercial imperatives of the NOCs. During the 1980s and 1990s this dynamic was not considered especially problematic from the perspective of the Chinese leadership, as energy policy was not a priority at that time, and simply expanding oil supply through domestic exploration and production was sufficient to meet China's growing oil requirements. However, the costs of fragmentation in the oil sector became more apparent once China's oil import dependency started to skyrocket in the early 2000s, and other energy-related crises also transpired. The loss of administrative capacity to govern the oil industry reduced the central party-state's capacity to formulate and implement more coordinated, comprehensive, sophisticated and coherent policy responses to address these energy challenges. Hence from 2003 onwards the oil sector's governing apparatus became the subject of capacity-building efforts aimed at reconstituting hierarchical political authority with a focus on formulating more effective energy policies.

6
Rebuilding Oil State Capacity (2003–2013)

China's shift to the status of net oil import in 1993 began to steadily push oil security up the Chinese government's policy agenda throughout the 1990s (Andrews-Speed 2010: 32). However, the continuing ability of international oil markets to meet China's oil demand during this time served to lessen the perceived insecurity caused by growing oil import dependency (Andrews-Speed 2010: 32). Beijing's attitude changed dramatically in 2003 in response to a range of energy security challenges that precipitated a sense of energy crisis in China. Internationally, the Iraq War and the subsequent increase in United States' active presence in the Middle East "reshaped China's basic conception of the geopolitics of oil and added urgency to its mission to lessen dependence on Middle East supplies" (Goodman 2005). At the same time, China's booming economy led to a surge in energy demand, with the ever-widening gap between domestic oil production and consumption needing to be filled by imported oil. This prompted the securitisation of energy issues in China, and pointed to the need for the Chinese government to improve oil policy coherence and coordination for the purpose of securing foreign supply, as well as tackle demand-side initiatives aimed at reducing domestic oil consumption. Domestically, China suffered an energy crisis from late 2002 to 2005, where the country experienced widespread power shortages, which led to a marked increase in oil demand as diesel generators were run to maintain power for industrial enterprises (Yergin 2011: 210; Downs 2006: 6). The only immediate alternative to coal in order to satisfy the shortfall in supply for power generation was oil, which explains why China's oil demand in 2004 jumped by 16 per cent, much more than the anticipated 7 to 8 per cent, causing a rapid increase in oil imports (Yergin 2011: 210).

Flaws in the extant institutional arrangements for governing the power industry, characterised by fragmented and unclear lines of authority (as a result of the decentralisation of political authority that occurred in the first Reform Era in China), were considered largely responsible for the power shortages and subsequent surge in oil demand (*China Daily* 2003, 2004c, 2004d and 2010b). In response the Chinese leadership pushed ahead with energy sector reform with renewed determination from 2003 onwards. Andrews-Speed (2010: 32 and 39) claims the shortfall in domestic energy supplies was the main catalyst for change in China's energy policymaking apparatus, where two key reform priorities came to the fore: "recentralise control over the energy sector and provide for more coherent policymaking". Kong (2006) argues that China's energy institutions have been the main source of energy insecurity, suggesting that these power shortages and other energy problems were caused primarily by institutional failings.

Throughout the late 1980s and 1990s the role of guaranteeing China's oil supply was left to the NOCs as detailed in Chapter 5. These state firms had become *de facto* leaders of the oil sector, both a consequence of decentralisation of political authority, and a legacy of the centrally planned system, wherein they had functioned as line ministries and continued to assume bureaucratic identities. CNPC in particular took the lead in oil policy formulation, which resulted in a narrow focus on oil supply, corresponding with its corporate strategic imperatives. This situation was increasingly viewed as insufficient to ensure the country's energy security in light of the alarmingly rapid growth in China's oil demand and the aforementioned domestic energy crises. In order to deal effectively with these energy challenges, China's policymakers began to consider initiatives aimed at actively securing oil supply and achieving greater efficiency and overall reductions in oil consumption, rather than simply continue to focus narrowly on expanding oil production (Andrews-Speed 2010: 32; Meidan et al. 2009: 609). The emergence of more complex and interlinked energy problems and their proposed solutions meant the central government could no longer afford to leave oil policy in the hands of the NOCs, and began to reclaim and recentralise oil policy authority. Numerous policy documents from the turn of the millennium to the present day reveal Beijing's preference for the defragmentation and recentralisation of the oil industry (and other energy sectors). Energy sector reform efforts since 2003 clearly have been oriented toward achieving this goal, and there has been much discussion on the apparent need for a unified energy ministry. Powerful institutions within the central party-state continued to oppose the creation of an energy ministry, notably the NDRC and state energy firms, such as the NOCs (Downs 2008a: 129; Yeo 2009a: 738). However, the view that an energy ministry is critically important for oil sector governance is in itself debateable, and very much informed by an FA perspective that emphasises weaknesses in bureaucratic structure as the primary obstacle to effective energy policymaking and oil industry reform. The BA perspective, on the other hand, emphasises those political, organisational and financial instruments that enable the central party-state to implement its policy agenda. From this perspective the lack of an energy ministry is not a major impediment to energy policymaking, as the party-state has other policy instruments at its disposal to implement its policy agendas.

While energy reforms were a response to a variety of energy challenges that emerged over the past decade, they have also been influenced by wider efforts to remake China's system of economic governance. China's ongoing transition from a planned to a market-oriented economy has seen the adoption of the regulatory state as an appropriate target model for effective economic regulation (see Yang 2004; Pearson 2007). The emergence of a fledgling regulatory state in China impacts governance of the oil sector in ways that the FA model cannot explain. The Chinese leadership has implemented several waves of administrative reform since 1993, which have been informed by a 'regulatory ethos'. In the western liberal economies regulatory reforms stress a reduction in open government intervention and the development and maintenance of an arm's length relationship between the state and market. However, China's regulatory system operates within a particular set of political and economic relations that at this stage of transition inhibits genuine independence of the new regulators (Pearson 2007). Nonetheless, China's regulatory reforms have changed the ways in which government agencies behave. Under China's regulatory system SOEs no longer function simply as agents of the plan, rather they are expected to be more profit-oriented and internationally competitive. Beijing's mission to rehabilitate an unprofitable state sector by implementing the 'grasp the large, release the small' policy (*zhuada fangxiao*), and to corporatise large SOEs with the aim of turning some of them into 'national champions', has so far met with reasonable success. For instance, in 1997 (around the time the 'grasp the large, release the small' policy began to be implemented), the entire state sector in China made total profits below 0.6 per cent of GDP, but in 2007 state sector profits accounted for 4.2 per cent of GDP

(Naughton 2008b: 19). Moreover, in 2010 three Chinese state enterprises (Sinopec, CNPC and State Grid) broke into the Fortune 500's top ten list (*China Daily* 2010c).

Since oil is a strategic or lifeline industry, linked to every other sector of the Chinese economy, it remains closely controlled and monitored by institutions at the apex of the party-state. This further compromises the independence and authority of the new regulators within this particular policy sector. Indeed this shows that China's fledgling regulatory state has in some ways simply been grafted onto the rest of the system – "squeezed in, but not allowed sufficient authority" (Pearson 2007: 727). The discrepancy between institutional design and institutional practice with regard to the new regulators in China arises from "the fact that the Chinese leadership's main concern is not how to make market principles take root in the domestic economy, but how best to deal with crucial state assets by expanding and strengthening firms (*you da you qiang*) on the global stage" (Yeo 2009b: 1032). An example of this can be found in the activities of the NDRC, which is the most powerful state planning agency, formally responsible for macroeconomic management, and has been given responsibilities usually left to firms and regulators in market economies, such as 'approval of large investment projects proposed by state firms and oversight of pricing in energy and infrastructure sectors' (OECD 2009: 96). Moreover, the CCP has retained ultimate control of the strategic industries through the tools of small leading groups, party committees and the nomenklatura system. While the party-state may undermine the independence and authority of regulatory agencies by intervening in the strategic sectors through the use of these mechanisms, it is also "the critical unifying factor in counteracting fragmentation" (Andrews-Speed 2010: 22). Hence it is important to understand developments in economic governance that have occurred during the second Reform Era in order to explain the ways in which China now manages the strategic sectors of its economy, with a mix of market incentives and interventionist modes of governance. The establishment of SASAC under the State Council in 2003 strengthened the strategic coherence of the reform and management of state firms, and ensured that "both the short-term and their long-term development strategies are in line with the interests of the Chinese state" (Kong 2010: 27). The oil sector remains state-led, and has become more centrally coordinated as policymaking authority continues to be reclaimed and reorganised by the central party-state.

Within China's oil sector the flow of political authority is generally hierarchical and top-down, and this has especially been the case during the second era of reform, since extant levers of party-state control have been strengthened and expanded, and more recent efforts to defragment and recentralise political authority have been undertaken. At the same time as becoming stronger and more centralised, top-down authority in the strategic sectors has also become more sophisticated and effective, for example, the performances of SOEs have improved through the use of new institutions of economic governance to enforce regulation and enhance competition and efficiency. When it comes to economic management in the strategic sectors the central party-state has thus far been able to strike a balance between maintaining state control and introducing market characteristics and other incentives to improve performances. Hence this chapter argues that the central party-state has improved its capacity to build more effective institutions to govern the oil sector. In the face of bureaucratic resistance the centre has demonstrated its ability to push through reform initiatives and other policy agendas, especially under conditions of crisis or where the policy initiative has key strategic or economic significance (Andrews-Speed 2010: 27). There has also been an improvement in the China's oil policymaking not only in response to crisis, but also in meeting daily oil requirements, though there remain some policy coordination problems that are explained by the FA model. These arguments runs counter to the predominant scholarly view that Beijing's slow moves towards recentralising party-state control of the oil sector are indicative of a patent inability on the part of the central party-state to build capable institutions and overcome bureaucratic opposition (Kong 2006; Downs 2004a; Meidan et al. 2009).

Furthermore, the nature of government reform, which has occurred in several waves over the past two decades at both the level of the national economy and in various industries, shows that the Chinese leadership usually favours gradual, incremental and cautious methods in achieving reform goals, rather than imposing abrupt systemic changes, which are potentially destabilising. This is largely reflective of a political culture that places a premium on harmony, stability and order. Here the *modus operandi* has traditionally been geared towards consensus and incremental transition, allowing new rules and institutions to be internalised gradually, and problems to be ironed out along the way. In fact greater legitimacy is derived from these methods, which have been a distinctive feature of bureaucracy in China for centuries. With regard to the oil industry, the institutional changes we have seen reflect a slow migration towards the ultimate goal of a centralised authority, rather than its achievement in one fell swoop. That being said, the central party-state has shown it can respond quickly and effectively in times of crisis, but otherwise its preference is to pursue gradual reform of extant bureaucracies and administrative apparatus where problems and conflicts can be resolved and the necessary expertise and capacity is built.

If one looks at the oil industry reform period from 2003–2013, we have seen a gradual evolution from fragmented political authority among multiple government departments and agencies to the creation of the ministerial-ranked NEC in 2010, and its administrative body charged with implementing energy policies, the NEA, which was formally established two years earlier and placed under the jurisdiction of the NDRC. It is likely that both these entities will eventually be merged into a single energy ministry at some point, but for now they also share authority for the energy sectors with the NDRC. The fact that this has not been achieved in a single round of administrative reform is by no means an indicator of state incapacity in the face of powerful vested interests. Rather it is more reflective of a distinctive political culture and preferred method for achieving reform (incremental transitions) that is repeated across various sectors of the Chinese economy. This is certainly not to say that FA is no longer a feature of the country's energy bureaucracy, but rather that it has been reduced and can be overcome when sufficient political will exists. Furthermore, the NOCs have been transformed into internationally competitive entities, whose interests by and large tend to dovetail with those of the Chinese state, since their commercial activities increase China's supply of oil. Where divergences of commercial and national interest have emerged, the central party-state has demonstrated its ultimate strategic control of these state firms. The NOCs may influence or advise the Chinese leadership, since they possess the necessary expertise and many of the heads of these companies were former petroleum bureaucrats with strong ties to the upper echelons of the party-state, suggesting that there are some important middle-up characteristics that impact oil policymaking in China. However, the strong centralising dynamic and control by the country's political elites indicates that ultimately political authority remains a decidedly top-down affair. This relationship between the central government and the NOCs is explored in detail in Chapter 7.

This chapter analyses reform of the oil governance regime since 2003. The first section addresses changes to economic governance at the national level, which has shaped the country's energy policymaking apparatus. This requires a brief return to 1998 in order to look at the major reforms of key government institutions and state industries that were undertaken at this time. This particular round of bureaucratic restructuring provides a compelling example of the party leadership's capacity to reach inside the Chinese state and reorganise the institutional apparatus. The 1998 administrative reforms affected most industrial sectors of the Chinese economy. They had a profound impact on the oil sector in terms of removing the remaining government functions from the NOCs and further preparing these companies for international competition. Following on from this, the Chinese leadership's attempts to recentralise the oil sector in response to specific energy challenges that have emerged over the past decade is examined. It will be shown that the central government has adopted a method of gradual transition toward unifying oil sector administration under a more centralised energy authority in order to improve the quality of oil policy formulation and implementation. The policy implications of this, and more importantly of the change in policy priorities of the Chinese leadership with a greater focus on energy security, can be seen in the shift from oil supply-side bias and the pursuit of *ad hoc* and sometimes contradictory initiatives, to a more comprehensive approach that takes into account demand-side management.

Government restructuring since 1998: Building the regulatory state

The majority of studies that deal with the policy process in China either fail to take account of the extensive political and bureaucratic reform efforts undertaken by the central party-state in order to improve state capacity (both within the CCP and the state apparatus) over the past two decades, or else acknowledge these reforms to some degree but underestimate their significance and even discount them as largely ineffectual (Shambaugh 2008: 3). Such reform efforts were a response to the excessive liberalisation and decentralisation of the 'free wheeling' 1980s in China, which culminated in a crisis of political legitimacy and authority in 1989. Furthermore, up until the early 1990s there was also an absence of a clear strategic plan for economic reforms at the national level. The *ad hoc* and experimental manner in which Chinese reforms proceeded during the first Reform Era was abandoned in favour of the development of a systematic reform agenda based on a vision for a new economic order during the second Reform Era (Yang 2004: 7; Naughton 2008a). Hence the reforms carried out by the third generation leadership led by Jiang Zemin marked a significant break with Deng's reform legacy, since it focused more on rebuilding or remaking the Chinese state to better support China's market-oriented economy. This presents a clear case of the party leadership significantly changing the institutional arrangements of the Chinese state, which is something that the FA model would deem unlikely to occur in China's political system.

The decade of intense introspection that followed the Tiananmen protests and collapse of the Soviet Bloc saw a reorientation of party-state priorities towards recentralisation and the development of a more integrated and flexible party-state apparatus. A more coherent and coordinated strategy with regard to market transition emerged; a significant component of which was the creation of a 'modern enterprise system', involving the conversion of SOEs into business corporations, predominantly taking the forms of limited liability companies (LLCs) and joint stock companies. This in turn created demand for legal and regulatory frameworks 'consistent with a rules-based environment' (Tenev and Zhang 2002: 16). In terms of industrial reform a particularly important governance strategy introduced in 1996, and enshrined in the Ninth Five Year Plan (1996–2000) was the policy of 'grasp the large, release the small'. This policy was driven by the central party-state's desire to tighten its grip over certain parts of the Chinese economy while withdrawing from other parts. To this end, the government typically targeted large SOEs in strategic sectors for further reform and support, with the aim of fostering 'national champions', able to compete in global markets. Small- and medium-sized assets were variously shutdown, handed to local governments or privatised (Wildau 2008: 28). China's efforts to rationalise, downsize and corporatise the state sector was a response to the massive loss-making situation of state firms, which had emerged by the early 1990s (Yang 2004: 31–32). Furthermore, whilst China protected itself against the worst effects of the Asian Financial Crisis, the country did experience a decline in economic growth, which prompted the 'grasp the large, release the small' policy to be adopted as official government policy at the Fifteenth Party Congress in 1997. In the wake of the Asian Financial Crisis, China sought to speed up the pace of administrative reform to strengthen the Chinese state, and more specifically to improve SOE performance, and improve the banking sector and other financial institutions (Morgan 2003: 344).

These developments constitute a significant component of larger efforts to remake China's system of economic governance to support an increasingly market-oriented economy (Pearson 2007: 720). Yang

(2004: 9) notes that special challenges arise from economic reform in socialist economies led by Leninist parties stating, "Rather than building state capacity to take care of the markets, the daunting task of market reforms in socialist economies not only involves the introduction of markets but also the rebuilding of the state into one that is qualitatively different and suited to markets." In other words, state institutions need to be reconfigured in order to support market reforms. In particular, China's leaders perceived the need to construct a regulatory system to enable better implementation and enforcement of economic regulation. By the mid-1990s the strains arising from the *ad hoc* nature of the 'growing out of the plan' strategy, which saw limited institutional changes to accompany the profound changes in the real economy, prompted this fundamental shift (OECD 2009: 38). Decentralisation had been a key part of the first era of reform, the motivation for which was of course the need to introduce market characteristics and incentives into China's economic system (Naughton 2006: 101). While China's economy had made remarkable progress under this strategy throughout the 1980s, growing imbalances and contradictions were becoming obstacles to further economic development (OECD 2009: 38).

Hence a major focus of the second Reform Era was on constructing the legal, regulatory and institutional underpinnings for a market economy (OECD 2009: 41). Pearson (2007: 721) claims that the Chinese government became aware of, and subsequently adapted, the independent regulator model through its contacts with international organisations such as the World Bank, OECD, Asian Development Bank and the WTO. Ultimately this new approach to economic management is underwritten by a fundamental shift in economic philosophy whereby the Chinese government has to some extent transformed its functions "such that it would desist from micromanaging firms and focus on providing macroeconomic control through economic levers" (Yang 2004: 8). Though as argued by Yeo (2009a), the central party-state still conducts micro-level interventions in the state sector, and this is particularly evident in some of the activities undertaken by the NDRC. Pearson identifies two main 'prongs' in the government's reform efforts towards the strategic sectors of China's economy. First was the corporatisation trend where SOEs were no longer controlled by line ministries and experienced some diversification in ownership (stopping well short of privatisation) (Pearson 2007: 719). Second, the state institutions responsible for governing the economy were restructured and upgraded (Pearson 2007: 719). Industrial ministries were either eliminated or downgraded, and ostensibly arm's-length regulators were also introduced (Pearson 2007: 719). Hence SOEs would cease to serve as administrative appendages of the government bureaucracy, and become more market-oriented (Yang 2004: 8). Pearson (2007: 719) claims that at the same time "parallel moves have been made to institutionalise stronger central authority over markets and to deepen economic supervision of the strategic sectors." In these strategic sectors the independent functions of China's regulators are constrained by party-state agencies and commissions. While China's regulatory regime ostensibly encourages market competition in the strategic sectors, in practice Beijing favours oligopolistic market structures in order to limit and manage competition among the large SOEs.

Economic reform since the mid-1990s in China has seen vast improvements in the capacity of the central government to effectively govern the market economy. Before 1995 Beijing possessed the view that the state had to exert at least 'nominal direct control' over all parts of the economy in order to maintain its ability to steer development. Since then it has adopted the idea that the government can exert effective control of the economy even if most output is produced in the non-state sector – "What matters is not the state's share of total output but which assets it controls" (Wildau 2008: 28). Since the oil industry is a vital strategic sector of the state-owned economy in China, the oil governance regime has been a locus of policy initiatives aimed at recentralisation, improving efficiency and policy coordination, and re-institutionalising central party-state control of energy policy formulation and implementation. Although the impetus for such recentralisation within the oil sector was a response to problems specifically associated with China's energy institutions, it also followed overarching institutional changes and reform drives initiated by the party leadership at the national level. Since 1993 there have been three waves of administrative reform geared towards the establishment of improved economic governance regimes that rely on regulatory systems; in 1998, 2003 and 2008. The 1998 reforms served to promote unity of administrative authority and curb bureaucratic fragmentation. In this round of bureaucratic restructuring the number of central ministries was reduced from forty to twenty-nine. The industrial ministries, which had been bulwarks of the centrally planned economy, were significantly affected by these reforms. The ministries were streamlined under the supervision of a powerful supra-regulatory body known as the SETC, which was created in 1993 and granted more far-reaching responsibilities in 1998. This commission was placed in charge of overseeing the operations of SOEs, which was an important development as the previous structure had given each ministry informal veto power in economic policymaking, often resulting in bureaucratic inertia (OECD 2009: 92). Hence another key tenet of this restructuring program was to centralise the supervision of a limited number (500–1,000) large enterprise groups under the SETC (Fewsmith 2008a: 210).

In order to keep tabs on the enterprises Premier Zhu Rongji established inspection groups, each headed by a person of vice-ministerial rank, who would report directly to the SETC (Fewsmith 2008a: 210). Fewsmith (2008a: 210) states "because [the inspection groups] were no longer ministerially based, they were expected to provide the government with more objective information on the state of the industry". The SETC was generally responsible for macroeconomic steering, for instance, through industry policy implementation. The SPC was renamed the SDPC (another supra-regulatory body later renamed the NDRC) in the 1998 reforms, and its main duties were forecasting medium- and long-range growth targets, maintaining an overall economic balance, curbing inflation and optimising economic structure (Yang 2004: 40). The SDPC's transformation into a regulatory authority did not quite materialise, as it has not been able to shed its "planned economy mentality": "[The NDRC's] mismanagement of inflation and energy policies prove that the 1998 reform failed to transform the government into a macroeconomic supervisor; the NDRC has been blamed for not only its continued micro-control, but also the reluctance to undertake root-and-branch market reforms in many sectors of the economy" (Yeo 2009b: 734). This major reorganisation of the state apparatus constituted the largest readjustment of the government-enterprise relationship since the beginning of reform in China.

Neither the SETC nor SDPC were authorised to directly supervise and intervene in the operations of the SOEs, hence granting the SOEs a higher degree of autonomy. The MOF and the PBC joined the SETC and SDPC as the central institutions of governance. These macroeconomic state agencies largely phased out micromanagement of SOEs, and instead concentrated on long-term guidance of industrial development (Pearson 2007: 721). While there remained some overlapping functions between the SDPC and the SETC, overall the 1998 government reforms significantly reduced the institutional overlap that could induce bureaucratic deadlock (Yang 2004: 56). Yang (2004: 41) claims that despite the strong legislative mandate for these reforms, there was much scepticism surrounding their implementation, as "the history of Chinese bureaucratic reforms is littered with aborted reform plans".

However, the streamlining of the central government proceeded very quickly, confounding the sceptics, and indicating that where sufficient political will exists, the Chinese leadership can readily overcome bureaucratic opposition and push through reform swiftly and effectively (Yang 2004: 41–42). Andrews-Speed (2010: 39) offers an FA view of the reforms, contending that with authority split between the SETC and SDPC, the "degree of coherence in energy policymaking deteriorated rather than increased, not least because of bureaucratic competition". Certainly the lines of authority between the SETC and SDPC were not always clear and subject to functional overlap. But while the 1998 reforms did not solve the problem of FA throughout the major policy sectors, they did attempt to address it.

The next wave of reform occurred in 2003 and expressly dealt with the problem of how to "promote the trend toward a relatively neutral regulatory state and yet maintain proper and efficient supervision over the multitude of state enterprises" (OECD 2009: 93). To this end, the SETC was dismantled and its bureaus on SOEs were transferred to the newly created SASAC. Parts of the MOF, namely those relating to equity ownership and approval of equity transactions, were also merged into SASAC (Yang 2004: 61). This agency is ministry-level, operates directly under the State Council and was headed by Li Rongrong (former head of SETC) up until 2010 when he retired and was replaced by Wang Yong, who was subsequently replaced by former oil executive Jiang Jiemin in 2013. Jiang has since been removed from this post as corruption charges were brought against him in September, and a new replacement, Zhang Yi, was announced in December 2013 (*China Daily* 2010a; *China Daily* 2013; Li 2013). SASAC's mandate is to promote the strategic restructuring of SOEs and further separate government ownership, enterprise and management. The creation of this regulatory commission, which operates under the State Council, is an important component of the 'grasp the large, release the small' policy, as its main task is "to build large, successful state-owned enterprise groups able to dominate their sectors" (Pearson 2007: 720). Naughton (2010: 441) claims that SASAC's creation in 2003 "is a convenient milestone to date the beginning of the current system. SASAC's establishment did not mark any rupture with previous policy...but it provided a formal structure where one had been lacking, and created a bureaucratic interest that articulates policies and projects a more-or-less coherent vision of the future". SASAC "nominally exercises the government's role as owner" of SOEs, but the CCP continues to make the most important personnel decisions through the COD (Naughton

2010: 456). The remainder of SETC's functions relating to industry; industrial planning and policy, economic operations and control, supervision of investment in technical renovation, macroeconomic policy guidance on enterprises of all ownership types, promotion of small and medium-sized enterprises, and planning for import and export of raw materials – were handed back to the SDPC, which was subsequently rechristened the NDRC. These reforms left the NDRC with two particularly important functions; approving large investment projects by SOEs, and overseeing pricing in the infrastructure sectors (OECD 2009: 604; Yeo 2009). The removal of 'planning' from its name is indicative of the trend toward using "market-oriented mechanisms to manage the economy rather than reliance on approvals, permits and microeconomic interventions", and meant to signal the shift to macroeconomic regulation (Yang 2004: 62). However, as stated by Yeo (2009a: 742) the NDRC in many ways remains a planning agency given its continuing use of "powerful levers of micro-management, such as investment endorsement".

The most recent round of restructuring reforms, starting in 2008, is referred to as the 'super-ministry' reform (*dabuzhi gaige*). These reforms focused on two key issues: (1) improving macroeconomic regulation of strategic industries, such as energy, and (2) the construction of a social security system (Yeo 2009a: 731). The latter goal distinguishes the 2008 scheme from previous reforms, and is a key component of the Hu-Wen leadership's vision of a 'harmonious society' based on greater social equity and justice. The Chinese leadership's recognition of the importance of social security added impetus to state efforts to deal with a variety of social challenges, such as pollution, housing and the growing rural-urban divide (Yeo 2009a: 731). According to Yeo (2009a: 730 and 742) "the 2008 reshuffling highlights a strategically articulated state engagement for further industrial development, rather than aloof market-oriented oversight," which is consistent with the view of senior Chinese officials in the Development Research Centre of the State Council (interviewed by Yeo (2009a: 742)) that "What China now needs is further industrialisation with strong government support and plan, not market-oriented rules and institutions." This reform drive entailed the establishment of five super-ministries – industry and information, human resources and social security, environmental protection, housing and urban-rural construction, and transport, plus a ministerial-level energy commission (*China Daily* 2008). Further reform was slated for key institutions, including NDRC, MOF and the PBC, with the aim of enhancing macroeconomic control (Yeo 2009a: 731).

The creation of the super-ministries entailed consolidations of extant government agencies, ministries and commissions that had similar functions (Yeo 2009a: 731). Hence the period from 1998–2008 has certainly reshaped key elements of the state apparatus. The drive to recentralise political authority, ownership and control, and abolish the industrial ministries (the core of the planned economy), was not just an end in itself but also a means by which a more streamlined regulatory state could emerge. SASAC was particularly instrumental in achieving this, and has already shown success in producing internationally competitive SOEs and assisting the state sector's return to profitability. As Pearson (2007: 720) concisely states:

> The goal of state sector transformation, then, is not simply to reduce state control. Rather, the message to be taken from the accumulation of new organisational forms – corporatised firms, the SASAC, enterprise groups, and aspiring national champions – is the desire to make state control more efficient and state enterprises wealthier and more effective at carrying out parallel imperatives of the party-state.

However, due to the central party-state's continuing desire to control the strategic parts of the Chinese economy, micro-control also remains a dominant feature of economic governance within the state sector. The central party-state has maintained its firm (interventionist) grip on economic governance where desired, not only through mechanisms of ownership, personnel appointments, high-level oversight, and five year plans, but also through structuring markets in order to shape the nature of competition among SOEs.

Within the strategic industries the party-state has sought to strike a balance between controlling and protecting its financial and social interests in the large SOEs, and establishing "managed competition" in order to help state firms become more efficient, profitable and internationally competitive. This has been achieved partly through the NDRC's control of pricing in order to avoid excess price competition. More importantly, the government has determined the market structure within which these firms operate domestically by limiting the number of SOEs permitted to operate in a given strategic sector. In order to try and prevent the adverse economic effects of monopolistic market structures, such as inefficiency and lack of competitiveness, China has established oligopolistic competition in key sectors including the oil industry (Naughton 2008b: 127). Pearson (2007: 725) observes, "the trend has been to divide monopolies along business lines or into geographically defined regions, creating regional monopolies rather than head-to-head competition". This is indeed a far cry from free market competition, rather this sort of domestic market structure establishes 'limited and managed' competition (Naughton 2010: 444). Naughton (2010: 444) claims, "it appears to be sufficient to strip away the promise of a quiet life from state firms managers, and force them to improve their performance and results". Tellingly, when managers of state firms have become "too competitive", threatening sectoral profitability through "aggressive competition and price-cutting", the government has intervened to re-shuffle the top management among competing firms (Naughton 2010: 446). Naughton (2010: 446) states the reasoning behind such moves: "This gives managers a fresh perspective, dampening their enthusiasm for unbridled competition, and reminding them that the government is the ultimate owner of all the competing oligopolists." A notable example of this kind of government intervention occurred in the telecommunications sector in 2004, in response to aggressive price-cutting (Naughton 2010: 446). The COD announced without warning a reshuffle of the top executives at China Mobile, China Unicom and China Telecom (McGregor 2010: 84). Two of these Chinese companies were listed on foreign exchanges, and yet "the party did not even stop to think about the board and its legal responsibility for choosing the chairman and senior executives" (McGregor 2010: 85). A reshuffling of top executives also occurred in the oil industry in 2011 – a subject to be discussed further in Chapter 7.

While China's institutions of economic governance have been successful in producing more efficient and internationally competitive firms, they fall short of the benchmark regulatory model, as clearly they are subordinate to the political, economic and social imperatives of the central party-state. A number of China's powerful central agencies and commissions, notably the NDRC, are involved in regulatory matters and pursue key goals considered "potentially regressive from a pro-market point of view" (OECD 2009: 96). In particular the NDRC's powerful authority for investment endorsement for state firms, which induces state interventions in economic operations, precludes the complete transformation of government functions into macroeconomic regulators (Yeo 2009a: 730). The CCP also maintains its strategic and supervisory role in economic reform. Hence the party remains at the centre of a "network of interacting organisations that hold ultimate authority in the state-run corporate sector" (Naughton 2010: 457). In terms of how dynamic plays out in practice Naughton (2010: 457) claims:

> Ordinarily, the Party chooses to exercise power in the background, especially when the foreground institutions are running well, but in the event of a mishap, the direct control relationship can also be resumed. This is particularly useful in a situation where nascent regulatory institutions are imperfect, not arm's length, and sometimes incapable of performing their function. In these cases, the hierarchy simply makes rulings directly, especially through personnel decisions.

At this stage of China's economic development and transition from a planned to a market-oriented economy this mixed approach of market incentives and state intervention appears to have introduced the right amount of market discipline for the time being, and has produced generally positive results in terms of performance. These trends in administrative reform, which have served to redefine governance and the boundary between state and market in China and have also to some extent redefined the relationship between the central party-state and the SOEs, provides the necessary backcloth for understanding recent changes to China's oil governance regime.

Recentralising the oil governance regime

During the 1990s China's NOCs essentially led and managed the oil sector in the absence of major policy input from the Chinese leadership. The lack of policy or strategic direction emanating from the centre was partly due to diminished capacity resulting from institutional dismantling, decentralisation and fragmentation of the oil industry and energy governance regime in the 1980s. The fact that energy policy, and oil security in particular, did not command a great deal of attention during this time, played a big part as well in the lack of direction provided by the central party-state. China was also emerging from a thirty year period of oil self-sufficiency; the shift to net oil import dependency did not occur until 1993 and took the country by surprise (Downs 2006: 6). When Premier Zhu Rongji abolished the Ministry of Energy in 1993, just prior to China becoming a net oil importer, it was assumed that the country would continue to be self-sufficient in oil (Downs 2006: 6). As a result policymakers in China had not thought extensively about the vulnerabilities arising from dependency on foreign oil. Hence China's activities in the oil sector, both at home and abroad, during the 1990s reflected a narrow focus on oil supply, driven by the commercial activities and imperatives of the NOCs, without extensive government initiative or support. CNPC had effectively taken over governance of the oil sector in the absence of an energy ministry. This oil company was formed in 1988 from the upstream assets of the abolished MPI, and became the *de facto* centre of authority for the oil industry in China. CNPC's former president, Wang Tao (formerly the minister of the MPI), was particularly instrumental in providing strategic direction for the oil sector, which would later be adopted as official policy by the Chinese leadership. Kong (2010: 33) considers the NOCs to have functioned as "policy initiators" and "market regulators" during this time.

An example of how the NOCs functioned as policy initiators is CNPC's formulation of the 'go global' policy in the early 1990s (Kong 2010: 47). This policy was being implemented by CNPC before it became an official government policy. Although, it should be noted that this corporate strategy was consistent with earlier promulgations by the CCP concerning China's opening up to the outside world. Kong (2010: 38) claims that at the Third Plenum of the Twelfth CPC Central Committee convened on October 20, 1984, the principle of "utilising two types of resources, opening two types of markets, and mastering two types of skills" was adopted. Here the "two types of resources" and "two types of markets" referred to domestic and foreign resources, and domestic and foreign markets, respectively. Therefore CNPC worked within the strategic parameters already established by the CCP. CNPC's quest to pursue overseas investments was presented as a corporate strategy by Wang Tao in 1991 in response to declining domestic oil production and rapidly expanding domestic oil demand. The need to 'go international' was one of the three strategies of CNPC's overall development plan, soon to be followed by Sinopec and CNOOC (Kong 2010: 41). According to this 'transnational operations strategy' CNPC would "expand foreign economic and technological exchanges, increase foreign trade in various forms, open up to the international market, and grow China's petroleum industry through participating in international competition" (Kong 2010: 38). To aid this endeavour CNPC established its own international department to deal with oil exploration and production cooperation with foreign companies. The narrow focus on expanding oil supply was a consequence of leaving the oil industry's strategic direction up to the NOCs, where their commercial viability relied upon increasing oil production. There was little political will or incentive to formulate more comprehensive oil policies that would take into consideration demand-side approaches to the problem of energy security. This leads Kong (2010) to argue that China's domestic and international oil behaviour during the 1990s was not the product of top-down authority, but rather middle-up initiatives devised by the NOCs. The Chinese leadership willingly relinquished direct control of oil production and policy direction for two main reasons. First, as already mentioned, oil security was not considered a policy priority. Second, the top oil executives, such as Wang Tao, were former petroleum bureaucrats who "enjoyed the central leadership's trust and confidence" (Kong 2010: 43). The influence of the latter is important to note since the fluidity and crossover of personnel between the central party-state and the NOCs has seen the oil companies able to influence the policymaking process (Kong 2010: 23).

The decentralisation and corporatisation of the oil industry in the areas of production, prices and administration was initially driven by the need to increase oil production and the profitability of the oil industry. The last energy ministry was abolished in 1993, and in its place authority in the petroleum industry became divided horizontally among central bureaucracies. This produced a situation where too many actors at the central level were charged with governing authority over the petroleum industry (Kong 2006: 72). The three NOCs – CNPC, Sinopec and CNOOC – were placed under the direct supervision of the State Council following the abolition of the line ministries. The lack of an energy ministry to oversee the oil sector further enabled the NOCs strategic and operational autonomy. Since Wang Tao was the former minister of the abolished MPI, and in effect became the leader of the oil industry as the head of CNPC and a former petroleum bureaucrat with a ministerial rank. The presidents of both Sinopec and CNOOC had occupied lower bureaucratic positions than Wang Tao, and therefore took their cues from CNPC. The NOCs enjoy full and vice ministerial ranking (CNPC and Sinopec inherited ministerial rank and CNOOC vice ministerial rank), which up until recently placed them at an equal or higher footing than many of the government departments and agencies in charge of petroleum affairs. The growing importance of sufficient and secure oil supplies, especially since the early 2000s has also contributed to the growing prominence of these NOCs within the Chinese political system. In explaining the structure of the oil industry and how it was governed through the 1980s and 1990s the FA model is typically applied. Downs (2006: 16) concisely summarises this institutional context for energy policymaking, "The liberalisation and decentralisation of the energy sector that have accompanied China's shift from a planned to a market economy, along with bureaucratic restructuring over the past two decades, have resulted in a shift of power and resources away from the central government to the state-owned energy companies and in a fragmented institutional structure of authority over the energy sector." Since the 1980s there has been a procession of *ad hoc* agencies created to improve energy governance, and the ways in which authority has been split among various bureaucracies has continually changed.

China's oil policies from 2003 onwards gradually became more coordinated, coherent, assertive and comprehensive. A large part of the impetus behind improvements in the energy governance regime was the growing importance of the oil sector to China's overall energy security, and national security, equation. Several events in particular served to heighten the perceived insecurity derived from growing foreign oil dependency. The United States' military presence in Afghanistan and oil-rich Central Asia after the 9/11 terrorist attacks unsettled the Chinese government and drove fears that America's intervention and presence in this region could threaten China's access to oil resources. The United States' military occupation of Iraq two years later would preclude the development of oil projects there by Chinese NOCs, and this prompted a 'buying panic' on the part of these oil companies, which was enhanced further by subsequent overseas investment setbacks (Christoffersen 2005: 67). Events in Russia also drove Chinese concern over access to oil supply and fed the perception of a looming energy crisis in China. The case of Sino-Russian struggles over the Angarsk pipeline (from East Siberia to Daqing) is a prime example. Yukos (a private Russian oil company) had unveiled plans for the pipeline to go ahead in 2001, but these plans were suspended soon after due to the Kremlin's assault on this company (Helmer 2003). To the Chinese this came as a surprise – projected oil imports from this pipeline had already been written into China's Tenth Five-Year Plan (2001–2005). At the same time, Transneft, a Russian state pipeline monopoly, developed a proposal for a 'northern route' from Angarsk to the Pacific Ocean port of Nakhodka, which would be financially supported in part by Japan (Christoffersen 2005: 58). In March 2004 the Kremlin announced that it would favour the Japanese-backed project. This followed another Sino-Russian energy incident in 2002, whereby the Duma moved to prevent CNPC from acquiring Slavneft (Russia's seventh largest oil company), which led to China's withdrawal from the auction (*Stratfor* 2002). Other failures in China's efforts to invest overseas include CNPC's failure to acquire the North Caspian Sea Project in Kazakhstan in 2003 when ENI Agip, Shell and ExxonMobil exercised pre-emptive rights (Christoffersen 2005: 72). Hence the outbound mergers and acquisitions strategy conducted by China's NOCs faced a series of setbacks throughout the 2000s, another prominent example being CNOOC's failed bid to acquire the Californian oil company Unocal in 2005. These setbacks further reinforced a perception of energy insecurity and fed the Chinese leadership's belief that "the international oil market is not free and China's access to oil is not guaranteed through the market" (Kong 2005: 56). Oil security also became a high politics issue from 2003 when oil prices began to increase significantly, culminating in the oil shock of 2007–2008.

In light of these new and emerging energy challenges the State Council adopted the *Medium and Long-Term Energy Development Program from 2004 to 2020* in June 2004. This document outlined the need for Beijing to recentralise energy policymaking authority, and focus on improving energy security, the diversification of oil supply, regional energy cooperation, and the need to build a strategic oil reserve. The views presented in this program were informed by the securitisation of energy issues arising from the Iraq War in particular, and also from recent Sino-Russian oil debacles (Christoffersen 2005: 67). A major turning point in the development of China's energy bureaucracy had also occurred in late 2002 (Downs 2006: 6). Specifically the country suffered an energy crisis caused in part from poor domestic energy sector management. Starting in late-2002 electricity demand rose rapidly following a boom in heavy industry, which caused the energy intensity of the Chinese economy to rise for the first time in the Reform Era (Kennedy 2010: 144). From late 2002 to 2005 the country suffered power shortages, which contributed to a substantial increase in oil demand and imports as diesel generators were run to maintain power (Downs 2006: 6). Already mentioned, China's energy institutions were considered to be at fault. Zhao Jianping, a senior energy specialist of the World Bank in Beijing stated:

> The current energy shortage reflects the failure of the government's "command and control" approach to address energy sector issues... The government may need to rethink its long-term energy policy and strategy and redefine its roles... It should move from making project decisions to develop a coherent energy policy framework and create an enabling environment for companies to make investment decisions (*China Daily* 2004d).

The 'unwieldy bureaucracy' already in place had difficulty coping with changing market demand, and Zhao claims that the 'mismatch in energy supply and demand', which led to the shortages, was the result of 'over-control' by understaffed, poorly funded and fragmented government institutions (*China Daily* 2004d).

Around this time Hu Jintao and Wen Jiabao took over the reins of power, bringing with them a new vision for China's development called the Scientific Development Concept, which is partly directed towards the construction of a 'conservation-minded society'. Government think-tanks and agencies were also engaged in a re-evaluation of China's energy policy. One of the first documents to contain a proposal to make energy efficiency a policy priority was the Development Research Center of the State Council's *China's National Energy Strategy and Policy* published in 2003. The NDRC's Energy Research Institute and the State Council's Development Research Center then conducted a joint study into China's energy strategies and policy regime. The report, *China National Energy Strategy and Policy 2020* was published in 2004, and focused on state capacity building to improve energy policy formulation and implementation. According to Christoffersen (2005: 73) the report cited "lack of comprehensive national strategies with legal authority, lack of scientific decision making, lack of law enforcement capacity for energy laws, and lack of policy coordination between oil, coal, electricity and nuclear" as being the main capacity problems in the energy realm. The role of the NOCs in steering oil policy was also considered problematic. As already mentioned, the main problem with relying solely on the NOCs to confront China's energy security problems at that time was their narrow focus on expanding oil supply (Downs 2006: 2 and 25). Meidan et al. (2009: 609) and Christoffersen (2005: 73) note that the importance of demand-side measures has been evident in all energy policy documents since 2004. At the same time the NDRC issued their *Medium and Long Term Energy Conservation Plan*, which also stressed the importance of energy efficiency and conservation, and in addition provided specific targets and objectives (NDRC 2004; Meidan et al. 2009: 610).

In order to craft comprehensive oil policies that addressed demand-side management, the consensus among analysts was that a more unified and effective energy policymaking apparatus had to be rebuilt. Hence Beijing embarked on reform efforts aimed at improving the institutional capacity to deal effectively with a range of foreign and domestic energy security challenges. This involved the formulation of clearer policy guidelines, clarification of ownership and control (particularly of the NOCs) and a reduction in the number of bureaucratic actors participating in the policymaking and implementation process. Broader industrial policies framed these initiatives, the 'grasp the large, release the small' and 'go global' policies were important to the process of adapting and improving the overall effectiveness of the energy governance regime and specific policies aimed at securing oil production and supply to China. The 'grasp the large, release the small' policy saw the return to profitability of the state sector through the retention and reform of large SOEs in the strategic industries, which were restructured beginning in 1998 and placed under the regulatory authority of SASAC, established in 2003. Previously the NOCs had been nominally controlled by several bureaucratic agencies, making it more difficult for the government to influence their strategic direction.

Whilst the 1998 and 2003 reforms had established market-oriented institutions to steer industrial development, the oil industry was still largely left to its own devices in terms of specific policy direction. Although the SETC and SDPC were responsible for investment approvals and macro-regulation of the oil sector, they lacked "the necessary institutional expertise and human resources to conduct meaningful industry oversight" (Kong 2010: 16). In the face of increasing power shortages and concerns over oil security, an attempt to rectify this lack of specialist expertise and oversight occurred in March 2003 with the creation of the Energy Bureau under the NDRC. This bureau was widely considered to be a compromise among key stakeholders in China's energy bureaucracy. Downs (2006: 18) claims that the NDRC and the NOCs opposed the creation of an energy ministry to manage the oil sector (since both wished to retain their respective power over aspects of oil industry development). According to *China Daily* (2004a), one 'industrial source' claimed, "Such discussion cooled off after the top decision-makers suggested reform was so significant and complicated that they won't make it during the tenure of the current government." The tenure of that government ended in 2008, the same year a new round of deeper administrative reform occurred. According to Downs (2006: 18), "The ultimate establishment of the Energy Bureau under the NDRC served the interests of the NDRC and the energy companies. This configuration preserved the NDRC's influence and prevented the creation of another layer of authority over the energy companies."

The Energy Bureau's mandate was to formulate energy supply policy, but only in upstream oil exploration and production. Other NDRC departments had responsibilities for end-user pricing, energy efficiency and regulation of industrial sectors (IEA/OECD 2007: 270). Kong (2010:

17) observed further fragmentation of oil policy functions, noting that "authority for balancing the oil production market, setting petroleum prices, building refineries, drafting petroleum import and export plans, investing in the petroleum sector, and overseeing petroleum transportation all go beyond the Energy Bureau and is assigned to the Bureau of Economic Operations, the Department of Price, the Department of Industrial Policy, the Department of Trade, the Department of Fixed Assets Investment, and the Department of Transportation". The Energy Bureau was considered largely ineffective as it lacked capacity and political clout due to its inferior bureaucratic ranking, and the fact it still had to share policy authority with other bureaucracies. The bureau was also understaffed, consisting of only thirty positions in 2005. The number of staff in the Energy Bureau was later increased to fifty-seven, but this was still not sufficient to meet the organisation's responsibilities (Downs 2006: 18). Given the enormous mandate to be executed by so few staff, it was no surprise that the bureau was unable to develop comprehensive and effective energy policies. According to Zhu Chengzhang, an energy expert, in practice the bureau also did not possess the authority to coordinate among more politically powerful stakeholders, and was overwhelmed with issues such as project approval, neglecting more important issues such as strategic planning (*China Daily* 2004a).

To better coordinate relations among bureaucracies dealing with oil policy, and to further recentralise authority, the State Council established the thirteen-member National Energy Leading Group (NELG), headed by Premier Wen Jiabao, in 2005. At the group's first meeting Wen said that "the NELG would be mainly in charge of energy strategy and major policies, the development and conservation of resources, energy security and emergency responses as well as energy cooperation with foreign parties" (*China Daily* 2005a). This Leading Group was a response to the NDRC's complaints concerning a lack of central leadership support for energy sector management (Downs 2006: 19). In terms of their general function, State Council leading groups coordinate policy implementation within the government (Miller 2008b: 11). They tend to be established when a consensus needs to be built on significant issues that cannot be resolved through interagency bargaining. As such the recommendations of leading groups are likely to have considerable influence on the policymaking process, since they typically represent the consensus of these various actors. In terms of content, leading groups issue guiding principles, rather than specific policies, regarding the direction that bureaucratic activity should take.

The State Council also created the NELG's executive agency, the Office of the NELG (ONELG), otherwise known as the State Energy Office (SEO), responsible "for undertaking the day-to-day work of the Leading Group and supervising the implementation of decisions made by the Leading Group; monitoring the status of energy security, conducting forecasting and early warning on major energy issues, and proposing countermeasures to the Leading Group; organising relevant departments to study energy strategies and plans; studying major policies related to energy development and conservation, energy security and emergency preparedness, and international energy cooperation; undertaking other tasks assigned by the State Council and the Leading Group" (ONELG/NDRC 2011). In terms of bureaucratic authority, the ONELG possessed the rank of vice-ministry, below that of the NDRC and the NOCs, but higher than the Energy Bureau, and consisted of twenty-four members, mostly at the level of provincial governors (Feng 2008). This led some commentators to conclude that it really just functioned as a "high-level research institute and advisory council", rather than a body with the ability to drive policymaking (Kong 2010: 18; Downs 2006: 21). The fact that the ONELG was headed by Ma Kai, who was also the minister of the NDRC, suggests at the very least, possible overlap between the NELG and the NDRC, and influence of the latter over the former (Kong 2010: 18; Downs 2006: 21).

However, as leading groups tend to wield a lot of informal power and authority in the Chinese political system, it is difficult to assess their influence by referring solely to ministerial rank. Certainly the composition of the NELG's membership leads one to be cautious in dismissing the influence of this leading group. In hindsight it appears that the NELG was essentially a transitional authority to pave the way to the creation of a more centralised and unified governing authority. Within three years of its creation, the NELG promulgated the *Medium and Long-Term Renewable Energy Development Plan*, participated in amending the *Medium and Long-Term Energy Development Plan (2004–2020)*, and initiated the drafting of China's first Energy Law (Feng 2008). It also pushed ahead energy-saving programs with some success (Feng 2008). Hence the scorecard for the NELG is not as dismal as critics such as Downs (2006) and Kong (2010) tend to insist. The founding of the NELG also signalled the rising importance of energy policy to China's leaders and their intention to formulate more effective energy policies. According to Zhang Jianyu, a scholar from Tsinghua University, "The government's decision to set up a leading group is a major step forward but is far from enough... We should have a cabinet ministry and invest more human resources in mapping out and realising an energy blueprint" (*China Daily* 2005b). While the conventional view holds that the NELG was yet another ineffectual bureaucratic agency, it can also be seen as a component of a longer-term strategy to incrementally centralise energy sector oversight and policy coordination.

The fact that a higher-level central authority, such as an energy ministry, was not established sooner might also be indicative of the earlier need to deliberately limit the number of voices represented and heard at the ministerial level. According to Downs (2006: 20), "A few Chinese and foreign energy experts based in Beijing have speculated that China's top leaders decided to form a leading group rather than a ministry for energy because they recognised that a ministry would likely become another layer of bureaucracy captured by vested interests." Hence the widespread assumption that an energy ministry is necessarily the best type of bureaucratic organisation to improve policymaking at this stage of the oil industry's development can be questioned. Arguably limiting genuine policy input to a few main actors such as the NDRC and the NOCs makes for less complicated and speedier policymaking processes, which the party-state leadership perceives as vital for such a strategically significant policy area. Adding another layer of bureaucracy would further complicate and perhaps dilute central directives regarding energy, whereas currently China's leaders can directly command the relevant state planning commissions and the NOCs to implement policy. Analysts such as Downs (2006 and 2008b), Yeo (2009a) and Kong (2006 and 2010) who lament the lack of an energy ministry to coordinate policy and reduce structural problems of bureaucratic infighting and overlap are still coming at the problem from an FA perspective, as Yeo (2009: 739a) states, "...a fragmented institutional energy sector required a comprehensive energy ministry to better coordinate competing bureaucratic interests at home, and to orchestrate overseas investments and secure the supply of petroleum while coping with the global energy security". From a BA perspective, the lack of an energy ministry is not particularly problematic, though it may be desirable in the long run as China's economy matures, market reforms deepen and state ownership is perhaps reduced. This is because it is the steep power gradient within the party-state and the centre's organisational, political and financial capacities that currently have a greater impact in shaping and influencing policy outcomes. This suits the imperatives of the CCP, arguably much more so than adding another bureaucratic agency.

Again, the Energy Bureau and NELG can alternatively be viewed as transitional authorities developed to suit the exigencies of those particular stages of energy development in a transition economy. Moving on, the State Council Information Office's 2007 white paper on energy indicated Beijing's desire to continue with energy reform:

> China has stepped up efforts in the reform of its energy management system, improved the national energy management system and decision-making mechanism, strengthened unified planning and coordination among state departments and local governments, and consolidated the state's overall planning and macro-control in the field of energy development, with the focus on changing functions, straightening out relations, optimizing the setup and raising efficiency, so as to form a management system that centralises control to an appropriate degree, divides work in a rational way, fosters scientific decision-making, and ensures smooth enforcement and effective oversight.

The Chinese leadership pressed ahead with the 2008 super-ministries initiative, which aimed to further improve central party-state control and overall efficiency in key areas of economic governance such as transportation, environment, industries and energy. Prior to the announcement of the restructuring plans at the Eleventh NPC, the idea of establishing an energy super-ministry was proposed. Those in favour argued an energy ministry would create a higher degree of policy coordination, secure China's energy supplies and create a more stable domestic environment. Initially the energy reform effort looked as if it might be ineffective. Rather than creating an energy ministry, as anticipated, the NEA was formed, possessing only vice-ministerial rank, and situated within the NDRC (Zhang and Lee 2008: 2). This would place it on an equal, if not lower, footing with the NOCs (in the case of CNPC), and hence in theory it would struggle to exert authority over them. The NEA operates under the State Council, essentially replacing the now defunct Energy Bureau, and absorbing other energy offices from NDRC, ONELG and the nuclear power administration of the Commission of Science Technology and Industry for National Defense (COSTIND) (Downs 2008b: 42). It has a total of nine departments, including an oil and gas department, with a staff of 112 people, and is responsible for the daily administration of a number of agencies and sub-groupings (Feng 2008). In terms of its mandate with regard to the oil sector it is responsible for managing the oil and gas industry, planning oil development, promoting industry reform and managing national and commercial oil reserves (Downs 2008b: 44). The NEA's director from 2008–2011, Zhang Guobao, stressed that the NEA would not be bogged down by project approvals, as was the case with the Energy Bureau, and instead will be able to focus its attention on researching and drafting long-term energy policies, and amending laws and regulations (Zhang and Lee 2008: 3). Whereas the Energy Bureau was only responsible for upstream oil exploration and production, the NEA has obtained administrative control over China's refining industry and strategic oil reserves (Zhang and Lee 2008: 3). Importantly, the NEA also has a department for energy cooperation, with the power to "negotiate and sign cooperative agreements and contracts with foreign governments and institutions" (Zhang and Lee 2008: 3). In addition to initiating international cooperation on energy, this department can also coordinate energy development overseas, and has the authority to approve significant international investment in oil, natural gas, coal, electricity and natural uranium (Feng 2008).

In presenting their more optimistic case for the NEA, Zhang and Lee (2008: 3) claim that the most significant development is "the integration of the NEA into the decision-making process for price adjustments to energy products such as electricity and petroleum." The NDRC still retains ultimate control over energy pricing, but the NEA has the authority to propose price adjustments subject to NDRC and State Council approval (Zhang and Lee 2008: 2; *China Economic Review* 2008). The NDRC is now required to consult with the NEA before adjusting energy prices, which indicates its ability to solely dictate pricing may have been somewhat diminished (Feng 2008). While it appeared to fall short of an energy ministry, the NEA has significantly more power and capacity than its predecessors. It possesses a vice-ministerial rank, higher than the Energy Bureau, but below that of some state energy firms, such as CNPC and Sinopec. In addition the NEA has its own independent Leading Party Member's group – an instrument of communication with the central party leadership that few vice-ministerial entities possess (*Stratfor* 2008a and 2008b).

The Eleventh NPC also announced that a National Energy Commission (NEC) was to replace the NELG as a high-level energy advisory and coordination body, but its creation in 2008 stalled, apparently due to bureaucratic resistance. Analysts such as Downs (2006) and Kong (2010) saw this as further evidence of the party leadership's inability to overcome a fragmented institutional structure captured by powerful vested interests, namely the NOCs and the NDRC. The argument here is that the NDRC is loath to relinquish its ultimate authority to set end-user oil prices, as this is a critical instrument of macroeconomic control, and the NOCs simply reject the idea of being subject to greater central oversight and control (Downs 2006). The creation of the NEC was shelved until 2010, and many commentators attributed this to further recalcitrance from the NDRC and the NOCs. However, it now appears that in addition to some bureaucratic resistance, the NEC was placed on the back burner due to a major shift in Beijing's policy priorities that had occurred by mid-2008. Previously, under conditions of high global oil prices and energy shortages, the impetus for reform of the energy bureaucracy was much stronger. When financial crisis erupted in the United States and the credit crunch brought international trade to a halt, energy prices plummeted, and reform became the least of China's worries (*Stratfor* 2010). Once the financial crisis began to recede, Beijing refocused its attention on energy reform. On 27 January 2010, the State Council announced the formal establishment of the NEC under the State Council, and to be led by Premier Wen Jiabao and Vice Premier Li Keqiang. Judging by the membership of the NEC (see Tables 6.1 and 6.2), the body is like a cabinet within the Cabinet (Bo 2010). The NEC is a planning and coordination body that was granted the bureaucratic rank of premier, therefore it is indeed a super-ministry that surpasses the official ranks of both the NDRC and the NOCs. The NEC is responsible for drafting national energy development plans, reviewing energy security and coordinating international cooperation (*China Daily* 2010b). The NEA now functions as the standing body of the NEC, assuming administrative and policymaking control of China's energy sectors. Hence the current framework for energy policymaking involves the NEC, NDRC and NEA; an arrangement that is generally understood to be a political compromise and alternative to a unified energy ministry, or as this volume proposes, another transitional body deemed to be appropriate at this stage of development.

The members of the commission (see Table 6.1 for the original membership, and Table 6.2 for an updated list) reflect Beijing's serious desire to improve energy policymaking. Lin Boqiang, director of the China Center for Energy Economics Research at Xiamen University said, "The establishment of the NEC shows the government has raised energy issues to an unprecedented level... Such a super-ministry, which centralises the powers of different ministries, can help China make better use of its energy resources" (*Xinhua* 2010a). The NEC's membership is an "all-star cast" of the most important and influential ministers from

Table 6.1 Members of the NEC in January 2010

Name	Title
Wen Jiabao (Chairman)	Premier
Li Keqiang (Vice Chairman)	Executive Vice Premier
You Quan	Deputy Secretary General of the State Council
Zhu Zhixin	Director of Central Finance General Office
Yang Jiechi	Minister of Foreign Affairs
Zhang Ping	Chairman of NDRC
Wan Gang	Minister of Science and Technology
Li Yizhong	Minister of Industry and Information
Geng Huichang	Minister of State Security
Xie Xuren	Minister of Finance
Xu Shaoshi	Minister of Land and Resources
Zhou Shengxian	Minister of Environmental Protection
Li Shenglin	Minister of Communication and Transport
Chen Lei	Minister of Water Resources
Chen Deming	Minister of Commerce
Zhou Xiaochuan	Governor of People's Bank of China
Li Rongrong	Chairman of SASAC
Xiao Jie	Chief of State Administration of Taxation
Luo Lin	State Administration of Work Safety
Liu Mingkang	Chairman of China Banking Regulatory Commission
Wang Xudong	Chairman of National Electricity Regulatory Commission
Zhang Qinsheng	Deputy Chief of General Staff Development
Zhang Guobao	Vice Chairman of NDRC and Director of the NEA

Source: Xinhua 2010a

the State Council such as that of the NDRC, MOF, and the MFA (Cai 2010). The NEC also features the head of China's Ministry of State Security and the Deputy-Chief of General Staff of the People's Liberation Army (PLA). The inclusion of military and foreign affairs figures in the commission is a clear indication that energy policy is now seen and treated as a key national security concern (Cai 2010). It shows a determination to better integrate foreign and domestic dimensions of energy policy. The NEC's membership also implies that "whoever opposes the NEC's decisions will be up against some of the most powerful people in China" (*Stratfor* 2010).

Since the beginning of reform three decades ago, centralised energy agencies responsible for energy policymaking have been short-lived, transitional in nature or have lacked the capacity to effectively coor-

Table 6.2 Members of the NEC as of September 2013

Name	Title
Li Keqiang (Chairman)	Premier
Zhange Gaoli (Vice Chairman)	Executive Vice Premier
Xiao Jie	Deputy Secretary General of the State Council
Liu He	Director of Central Finance General Office
Wang Yi	Minister of Foreign Affairs
Xu Shaoshi	Chairman of NDRC
Wan Gang	Minister of Science and Technology
Miao Xu	Minister of Industry and Information
Geng Huichang	Minister of State Security
Lou Jiwei	Minister of Finance
Jiang Daming	Minister of Land and Resources
Zhou Shengxian	Minister of Environmental Protection
Yang Chuantang	Minister of Communication and Transport
Chen Lei	Minister of Water Resources
Gao Hucheng	Minister of Commerce
Liu Shiyu	Deputy Governor of People's Bank of China
Jiang Jiemin	Chairman of SASAC (charged with corruption, new SASAC Chairman appointed: Zhang Yi)
Wang Jun	Chief of State Administration of Taxation
Wang Dongliang	State Administration of Work Safety
Shang Fulin	Chairman of China Banking Regulatory Commission
Wang Guanzhong	Deputy Chief of General Staff Development
Wu Xinxiong	Vice Chairman of NDRC and Director of the NEA

Source: CPC News 2013 and Xinhua 2013

dinate energy policies among bureaucratic agencies due to their low rank. From an FA perspective these bureaucratic deficiencies heavily impact the policymaking process and thwart the emergence of effective energy policies. On the other hand, a BA view privileges the role of elite power and authority in the Chinese political system in effecting policy outcomes. As such the lack of an energy ministry is not necessarily the major problem it is portrayed to be in the FA informed literature. Rather in the context of market transition we see incremental moves towards bureaucratic recentralisation in the oil sector (and other energy sectors) in a way that fulfils party-state imperatives. There has been a gradual evolution since 2003 toward the creation of a central energy authority with sufficient rank to grant it enough political clout to coordinate energy policies among bureaucratic agencies and powerful state energy firms. This has more or less occurred with the creation of the NEC and NEA, and reform on this front is likely to continue as market reforms deepen. Adherents of the FA perspective such as Downs (2008a and b) and Yeo (2009a) remain doubtful that these new energy policymaking arrangements constitute a substantial institutional change, yet alone improvement. Yeo (2009a: 742) claims that the new energy agencies "appear to be simply a new name for old agencies with similar responsibilities". Upon closer examination of these agencies, including their rank, composition and direct links to the party leadership, it becomes clear that they have strengthened the energy governance regime. How this plays out largely remains to be seen, but certainly the NEA and NEC at least have sufficient authority to coordinate oil policies. Zhang and Lee (2008: 4) are more optimistic about recent developments in Chinese energy reform stating, "with concentrated and expanded regulatory functions, increased institutional independence and credible political strength, the government may have finally created a legitimate foundation for the eventual transition to a full energy ministry and a brighter energy future". Given the nature of incremental political and economic transitions in China it is likely that in several more years these two bodies will be merged to create an energy ministry.

The development of comprehensive and effective energy policies, and the construction of an appropriate state apparatus to implement them, is a long-term undertaking in China. Beijing has pursued these goals through gradual administrative and regulatory reform, the original impetus for which was the rise to prominence of energy security on the agenda of the Chinese leadership as a result of the domestic energy crisis and the emergence of international energy security challenges. This method of gradualism has its own legitimacy within the Chinese political system, and has produced much better results for China's political and economic development than the shock therapy or big-bang approaches implemented with such disastrous consequences in other countries, notably Russia during the 1990s. Hence it comes as no surprise that the party leadership continues to adhere to incremental institutional change. Incremental change ensures stability, harmony, the maintenance of order as well as adequate time to learn and internalise new rules and norms. At the same time the central party-state has the authority and capacity to push through their policy agendas since they operate as the most powerful agents that are external to the state and able to reshape and reorganise the state apparatus to suit their interests. Changes to the energy governance regime have seen a shift from fragmented, bottom- and middle-up flows of political authority, to an increasingly coordinated and top-down mode of governance that relies upon both market-based and interventionist approaches. In the case of the oil industry, the Chinese leadership retains direct connections to the NOCs and the NDRC; a relationship which is not mediated by a layer of bureaucracy that would be introduced with the creation of an energy ministry. For now this arrangement suits the Chinese leadership, allowing them to pursue deeper structural reform with minimal consultation.

Oil policy formulation and implementation

The recentralisation of party-state authority and resources, and rebuilding of oil state capacity that has been undertaken since 2003 has improved energy policy content and direction, and policy outcomes for the oil industry. Within the oil industry, policymaking authority has been reclaimed by the central party-state, and is no longer solely determined by the NOCs. This has translated into a policy shift from a narrow focus on expanding oil production, to the development of more comprehensive policies that include a focus on demand-side management. Events that transpired at the turn of the twenty-first century, notably the Iraq War and a series of domestic energy crises, led former President Jiang Zemin and Premier Zhu Rongji to call upon the government to draft clearer and more coherent energy security strategies. Together the SETC and the State Development Planning Commission (SDPC/NDRC), worked with the party leadership to finalise China's oil policy approach from 2002–2003 (Kong 2010: 56). These initial efforts, which can also be located in the Tenth Five Year Plan, still form the backbone of present-day oil policy approaches in China. On 26 December 2007 the State Council Information Office published a white paper titled *China's Energy Conditions and Policies*. This white paper addresses China's energy circumstances and policies, with a focus on policy and planning for the future. The 2007 white paper was the first of its kind to provide a comprehensive overview of China's energy development and outline the country's energy policies with regard to energy "goals, supplies, efforts in conservation, development of new technologies and international cooperation" (State Council Information Office of China 2007). Rather than being devoted to policy and planning, two previous white papers released in 1995 and 1997 took the form of annual reports (Feng 2008). In addition, the

Eleventh and Twelfth Five Year Plans, and other energy policy documents produced by the State Council, detail China's energy policies. Hence over the past decade the attention devoted to energy security by the Chinese leadership has been unprecedented, at least during the Reform Era.

The Chinese government does not necessarily promulgate specific oil policies rather they are addressed within the broader context of energy policy, and follow the contours established with reference to energy more generally (though the majority of energy policy refer to measures relating to oil supply security). Oil supply security constitutes the central component of China's energy policy given the country's high level of import dependency, global crude oil price volatility, and so on. The Tenth Five Year Plan for the period 2001–2005 in particular forms the basis of the country's energy policy and has been repeated in various government documents, albeit in more refined and updated versions. Issued in 2000 the plan placed emphasis on the need to diversify the sources of oil supply, build a SPR and implement the 'go global' policy. In order to achieve these goals the NOCs were earmarked as the main actors in charge of implementing China's oil policies. In 2002 Chinese government agencies, namely the SETC and SDPC/NDRC drafted the country's oil security strategy. The oil security strategy stipulated that China's energy institutions had to be upgraded, and a State Energy Commission established to better coordinate energy policy implementation among the relevant bureaucracies. The Tenth Five Year Plan and oil policies outlined by the SETC and NDRC mainly provided an endorsement of extant policy approaches, such as the 'go global' investment strategy pioneered by the NOCs.

The Hu-Wen leadership was more proactive in dealing with the problem of energy security in a comprehensive fashion, addressing both supply and demand-side management. The Eleventh Five Year Plan for the period 2006–2010 presented a blueprint for the energy sectors stressing the development of a 'resource-efficient and environment friendly society'. In keeping with the theoretical framework for economic transformation devised by the fourth generation leadership, the Scientific Development Concept, the Eleventh Five Year Plan primarily emphasises energy efficiency and conservation. Ambitious and binding targets were set for reductions in energy intensity over the course of the plan: "Resources utilisation efficiency will be improved considerably while energy consumption of per unit GDP will be lowered by 20 per cent" (*China-Gov.cn* 2006). This drive for energy efficiency was mainly a response to the fact that domestic energy industries have been unable to keep up with soaring energy demand resulting from the surge in economic growth starting in the late 1990s; blackouts and fuel shortages had become widespread by the end of 2002. The plan also provides a means of addressing energy supply security, as the aim is to reduce energy use in future development. In terms of their impact on the oil sector, these policies reveal China's desire to start tackling demand-side management to enhance energy security by reducing overall oil consumption. Since the end of this Five Year Plan period has been reached, the outcomes can be evaluated. Government reports indicate that the country's energy intensity fell by 15.6 per cent from 2005 to 2009 but then rose slightly in 2010 (OECD 2010: 98). Although falling nearly 25 per cent short of the 20 per cent target, such gains realised over a short period of time are impressive nonetheless. However, Andrews-Speed (2009) tempers this view somewhat by suggesting that the primary reason behind success to date appears to be the government's decision to focus its efforts on those sectors that would yield the greatest short-term results, such as heavy industry.

These energy policy approaches were expanded upon in the Twelfth Five Year Plan for the period 2011–2015. This plan in particular shows that Chinese policymakers are clearly cognisant of the immense challenge for China in meeting such large-scale energy needs over the coming years and also in addressing severe air pollution (largely attributed to coal combustion). Under this economic plan the emphasis on breakneck economic growth is diminished in favour of sustainable growth and environmental protection, with a focus on clean energy development. In other words, China's economic development strategy places more emphasis on being "resource efficient and environmentally friendly" (Cheng 2008: 312). Emphasis was placed upon further improving energy efficiency and increasing the use of clean energy (OECD/IEA 2011: 78). The Twelfth Five Year Plan's energy program consists of several major initiatives. First, with regard to energy development in general, China will seek macro-level adjustments of both supply and demand, rather than simply focusing on energy supply (Jiang and Liu 2011). In achieving this, the plan places emphasis on technological innovation, gives priority to domestic energy production, encourages smart grid development, seeks to expand the country's energy transportation infrastructure and the SPR, encourages the construction of more oil and gas pipelines and promotes clean energy development concentrating on hydro, biomass, nuclear, solar and wind power (Jiang and Liu 2011). In addition to improving energy security, huge investment in energy-related projects, particularly clean energy development, is also aimed at helping China to achieve "independent innovation" and move beyond low-end manufacturing. The proportion of non-fossil fuels in primary energy consumption is targeted to reach 11.4 per cent by 2015, up from 8.3 per cent in 2010 (Jiang and Liu 2011). Beijing aims for clean energy sources to constitute 15 per cent of China's total energy use by 2020. The Twelfth Five Year Plan also outlined further targets for reductions in energy intensity and carbon dioxide emissions. Hence there is a clear desire to shift China's energy use away from reliance on coal and oil for both environmental and security reasons, that is, to address pollution concerns and reduce external dependence, particularly on foreign oil supply.

While securing oil supply continues to be the most important goal to achieve energy security, the Chinese leadership acknowledges that greater efficiency needs to be encouraged on the part of end-users of petroleum products, particularly industrial users, to eliminate waste and reduce overall oil demand and cost. Increasingly effective economic and political incentives have been developed to promote energy efficiency. For instance, officials who fail to meet energy targets can be denied annual rewards and promotion (again the nomenklatura system proves to be a simple and effective mechanism of control). Strikingly, China became the world's leading investor in renewable energy in 2009, outpacing both the US and EU combined (*People's Daily* 2010a). In 2010 alone China spent US$54.4 billion on clean energy development, and the draft *New Energy Industry Development Plan 2011–2020* indicates that the Chinese government intends to invest five trillion yuan (US$791 billion) on clean energy by 2020 (*People's Daily* 2010a). While success in renewable energy development cannot solely be measured by the size of investment, the rapid expansion of wind and solar industries in China is an achievement, though there are challenges in converting extant energy systems to clean energy (given the short period of time that has elapsed these may be addressed through further investment and reform) (see, for instance, Lu 2010). Even though big efficiency gains and clean energy developments are unlikely to significantly reduce oil consumption over the coming years, they certainly indicate that the centre has the capacity to implement its core policy objectives.

Focusing more specifically on oil, Beijing now pursues a multi-pronged and integrated approach to securing sources of oil supply abroad, combining energy security goals with foreign policy initiatives and oil diplomacy. This has helped China's NOCs to secure a large number of foreign oil deals in various regions of the world. China's oil import diversification strategy is another policy area in which improved policymaking and coordination has produced the desired results within a short time frame (Vivoda and Manicom 2011: 237). The country's oil volumes have grown by an average of 17 per cent annually, but rather than leading to increases in Middle Eastern volume share (as one would expect), import volumes from the Middle East have remained steady, and have instead increased from Africa, South and Central America and Central Asia (Vivoda and Manicom 2011: 236). China now imports oil from over thirty countries, which indicates a high level of diversification has been achieved in a short period of time (mainly over the past decade). The overall share of South and Central America, North Africa and West Africa has risen from just under 5 per cent in 2001 to over 27 per cent in 2009. Vivoda and Manicom (2011: 246–247) argue that three dimensions of China's diversification strategy help explain these impressive increases in oil import diversification: (1) Beijing's strategy of securing oil supply from pariah states that have been isolated by the international community, such as Sudan and Iran, and reluctance to rely solely on the sea lanes that are protected by the United States; (2) The implementation of China's 'go global' policy by the NOCs, which serves both commercial and strategic interests; (3) The fiscal capacity of the Chinese state – the NOCs receive extensive financial support and cheap loans, and Beijing also offers soft loans to oil-rich countries in order to sweeten oil deals with Chinese NOCs (Vivoda and Manicom 2011: 243). Clearly China appears willing to pay a premium for oil import security and has the capacity to do so (Vivoda and Manicom 2011: 243). This diversification strategy is also preoccupied with transportation security, another major focus for Beijing. China has conducted many studies into alternative oil supply routes that avoid strategic chokepoints such as the Strait of Malacca, and also the Strait of Hormuz (Erickson and Collins 2010: 90). As a result oil pipelines to China have been constructed, or are under construction, from Kazakhstan, Russia and Myanmar. In addition to actively exploring alternative supply routes, Beijing has also embarked on diplomacy that enhances relationships with "petroleum corridor countries", and has sought a greater role in protecting the Strait of Malacca. While scholars such as Erickson and Collins (2010: 92–93), debate the economic logic of some of these projects, and whether they actually enhance energy security (the argument can be made that pipelines are more vulnerable to disruption than seaborne delivery), these developments do indicate a growing institutional capacity in China to pursue a multidimensional and proactive approach to energy security.

Finally, efforts by the central party-state to produce globally competitive NOCs charged with implementing oil policies and ensuring stable, reliable and affordable oil supply has been largely successful. Considering their latecomer status to the global oil industry, China's NOCs have quickly climbed the competitiveness ladder. CNPC, Sinopec and CNOOC receive strong financial and political support from the Chinese government to undertake oil activities at home and abroad, and have been well managed under SASAC. The most significant element of oil policy implementation arguably hinges on successful coordination and synchronisation between Beijing and the NOCs. This relationship and the successful integration of China's energy security with foreign policy, constitutes the subject of the final case study chapter.

7
China's National Oil Companies 'Go Global'

China's large SOEs operate in the commanding heights of the economy, that is, mostly in the strategic sectors such as energy and power, industrial raw materials, military industry, large-scale machinery building, transport, banking and telecommunications. These SOEs are centrally-administered and guarantee China "a stable supply of energy and main industrial inputs such as steel and chemical materials, safe financial operations, a smoothly operating transportation system, fast-growing communication services across the country, continued progress in high-tech industries, and a substantial increase in automobile production" (Liu 2009: 546). The strategic sectors of the Chinese economy play a vital role not only in economic growth but also in economic transformation; they are frequently used by the central party-state in order to pursue a range of industrial, social and political objectives. Beijing's focus on reforming and growing the state sector has led scholars such as Huang (2008: 277) to conclude that China is now best characterised as a commanding heights economy, with private enterprise squeezed out of many industries that receive extensive government support. Once the state sector was effectively downsized following the implementation of the 'grasp the large, release the small' policy (*zhuada fangxiao*) during the Ninth Five Year Plan (1996–2000), the central party-state then focused on returning those remaining large state firms, dubbed 'national champions', to profitability. The primary executor of this agenda since 2003 has been SASAC. Through a range of policy initiatives, and under the ownership and regulatory supervision of SASAC, China's NOCs have been further transformed into internationally competitive state enterprises.

China's NOCs are new global players that now compete not only with each other internationally, but also with international oil companies (IOCs) and NOCs from other oil-consuming countries. Since China's NOCs retain close links with the central party-state their increasing international oil activities have been the subject of much scrutiny and conjecture over the past decade. Specifically, the nature of the relationship between these NOCs and the central party-state, and the ways in which they operate domestically and within international oil markets are now widely debated in policy, academic and business communities. The close and somewhat opaque nature of the Chinese government-NOC relationship has become a source of anxiety for other oil-consuming countries, other NOCs and IOCs, and, in some cases, the host countries of Chinese oil investment. The reasons for this concern vary from unfair competition due to the competitive advantage NOCs receive from Beijing's political and financial support, to anxieties over energy and national security (see Luft 2004; Jaffe and Lewis 2002). Within the extant scholarly literature opinion is polarised. At one end of the spectrum scholars argue that China's NOCs are government appendages subject to extensive control and direction by the central government as they execute national policy (*Knowledge@Wharton* 2012). In contrast, analysts such as Downs (2008a) and Houser (2008) see the NOCs as increasingly autonomous and difficult for Beijing to control. They argue that China's NOCs behave little differently to private oil companies and are driven by commercial imperatives, rather than national interest or other non-economic objectives. Li (2011: 26) claims that the strength of the NOCs *vis-à-vis* allegedly weak energy institutions in China has allowed these companies to influence energy policymaking stating, "Because of this, it is the corporate interests of the Chinese energy firms rather than the national interests of the Chinese state that drive China's energy agenda." From this perspective China's state-led oil strategy is deemed to be the product of bottom- or middle-up initiatives (driven by the NOCs), rather than top-down processes. On this point, Kong (2010: 27–28) agues that China's oil industry is characterised by a co-governance structure, where both the NOCs and the central government co-govern the oil industry, which seems to imply an equal power-sharing arrangement.

Arguably both perspectives regarding the relationship between the NOCs and the central government are one-sided and often based on a narrow conception of government control. The NOCs do indeed execute Beijing's energy security policies, but they are also massive bureaucracies that seek to defend their own interests, and constitute a significant power base and source of political patronage within the political system. However, in the end the central party-state retains the key levers of control and influence that can be deployed effectively to solicit compliance where the need arises. Furthermore the extant literature has largely failed to engage in sufficiently detailed examinations of corporate governance structures in China's NOCs and the domestic regulatory system under which they operate. Analysis of these institutions provides more compelling explanations of how the strategic sectors of China's economy are governed. A closer look reveals that China's NOCs operate under distinctive corporate governance structures where central government and CCP involvement remains pervasive. In this chapter corporate governance refers to the framework of institutions, laws, rules and policies that determine the way businesses are directed, regulated and controlled. Corporate governance is commonly defined by the interaction between shareholders and managers. In China's SOEs, corporate governance structures are the product of gradualist economic reform, whereby Beijing has sought to expose state firms to market forces so they can compete internationally, while at the same time retaining control.

In contrast to the conventional accounts, this study argues that the relationship between the central government and the NOCs can be characterised as 'collaboration governed by hierarchy'. The central party-state remains the pivotal player that establishes strategic direction for the industry, and has cultivated a range of policy instruments through which it can exert its authority and intervene, although it chooses to do so infrequently (Naughton 2010: 457). This argument recognises that China's NOCs possess valuable specialised knowledge of the oil industry. As such they fulfil an advisory capacity to influence energy policymakers, and possess operational autonomy to conduct their day-to-day business operations largely free from party-state intervention. Ultimately the central party-state wants the NOCs to be commercially-oriented and internationally competitive, but at the same time these companies are required to fulfil some non-commercial objectives. Herberg sums up this point by saying, "Today, you can [view] 80 per cent to 90 per cent of [the NOCs] behaviour to be what a company of that size in the industry would normally do to compete, and then there's that 10 per cent that has to be responsive to Beijing's politics" (quoted in *Knowledge@Wharton* 2012). Hence, it is important to understand at the outset in this chapter that the central party-state has little interest in micromanaging SOEs to any great extent – it largely divested itself of this role during the first Reform Era. Rather its main interest lies in macroeconomic management and steering, though it retains the capability to conduct micro-level interventions.

The central party-state has deliberately encouraged the development of the NOCs' operational and technical autonomy in order to promote efficiency and competitiveness to serve the economic needs of China and ensure the country's energy security. As a result China's NOCs behave similarly to private companies most of the time, especially on the international stage. However, the centre possesses several powerful political, organisational and financial levers that enable Beijing to intervene in the commercial operations of state firms, and also determine their strategic orientation and direct their foreign investment toward targeted regions and countries. Government intervention in corporate matters has occurred in the domestic sphere involving the parent or holding companies, whereas the international commercial activities of the publicly listed subsidiaries are subject to more indirect forms of control or interference. Due to the general convergence of corporate and national interests in the oil sector, state intervention in the NOCs' commercial operations does not occur often. In cases where political and commercial interests diverge, for example, when Sinopec reduced its refined oil output to stem financial losses during the international oil price hikes from 2005–2008, central party-state objectives are shown to trump those of state firms. Other examples include: the West-East Gas Pipeline Project (WEPP), efforts by the central party-state to prevent the accumulation of too much power among the top executives of the NOCs through reshuffling of company heads, and the political purge of several top oil executives on corruption charges (notably under the new leadership headed by Xi Jinping), partly for the purpose of removing obstacles to deepening state sector reforms.

This chapter is concerned with the nature of this relationship. In the following section the relationship between the government and the NOCs is evaluated by examining corporate governance practices in China's oil industry through ownership structures, senior management and boards of the NOCs, and the regulatory environment within which these companies operate. This will help to determine the degree of independence these companies have from the central party-state. In doing so, this chapter also discusses how the Chinese government has structured competition in the domestic market among the NOCs. Throughout impact of the NOCs' corporate governance upon their actual behaviour is addressed. In explaining the nature of the government-NOC relationship, this chapter also delves into how effective the NOCs have been in implementing the 'go global' (*zou chuqu*) policy in their hunt for foreign oil assets. The multifaceted approach Beijing takes to securing oil supply, which includes extensive financial support for the NOCs and oil diplomacy is also analysed. This state-led approach to securing foreign oil supply, which has notably entailed rapidly diversification of the sources of oil supply, appears to work reasonably well; Chinese NOCs have secured a large number of oil deals in a wide range of countries in various regions of the world in recent years.

Corporate governance with Chinese characteristics

Through a process of gradualist reform China's NOCs – CNOOC, CNPC and Sinopec – have been transformed into globally competitive SOEs with subsidiaries listed on domestic and international stock exchanges. Poor performance among Chinese SOEs provided the impetus for the latter phase of this transformation from the mid-1990s onwards, whereby the focus of ongoing enterprise reform shifted to ownership and corporate governance restructuring (Ewing 2005: 319; Naughton 2010; Wildau 2008: 28). Despite the corporatisation of the NOCs, the Chinese government retains mechanisms or tools of control not only over the holding companies, but also their publicly listed subsidiaries through majority share ownership and a range of distinctive corporate governance mechanisms, which taken together are sometimes referred to as 'corporate governance with Chinese characteristics' (Liu 2006: 418; Ewing 2005: 320). In assessing the nature and extent of control that the Chinese government continues to wield at the firm level it is necessary to examine the institutions and mechanisms employed by the central party-state to manage and govern the NOCs and their publicly traded subsidiaries.

Since Beijing considers oil a strategic industry that plays an integral role in China's economic growth and development, the oil industry is more closely controlled and nurtured by the central party-state than most other sectors of the Chinese economy. Within these strategic or pillar industries China has chosen to reform and grow large SOEs, dubbed 'national champions', providing them with protected market positions and extensive financial support. In return these companies are expected not only to flourish within their respective industries, but also to advance Beijing's strategic and political objectives. Since the 'grasp the large, release the small' policy was adopted in 1996 the Chinese government has sought to institutionalise stronger centralised authority over these strategic sectors, so as to tighten Beijing's grip on the top tier of Chinese industrial firms and at the same time introduce market-based incentives to improve economic performances (Pearson

2007: 719). Naughton (2010: 443) claims that state control of the commanding heights of the economy allows the Chinese leadership to achieve "social goals, while also protecting national security and preventing foreign dominance of the economy". The most important social responsibility that the NOCs must fulfil is to guarantee a stable oil supply. In doing so company profits may be sacrificed, for example, government regulated oil prices stabilise the market for refined oil and the NOCs, namely Sinopec (as the country's chief oil refining company), are forced to suffer downstream losses when the international price of crude rises. Geopolitical and strategic considerations also influence the foreign investments undertaken by the NOCs (Chen 2007). Hence the Chinese government is loath to relinquish control of the oil industry, especially at this stage of market transition since this sector plays such an important role in domestic economic and social stability. The central party-state appears willing to absorb the cost (and has the fiscal capacity to do so) of maintaining this stability, which sometimes results in profit and efficiency losses for the NOCs. In recent years Sinopec in particular has been partly compensated by the government for its massive refining losses.

The close identification of strategic sectors such as the oil industry with national security was highlighted recently by the sentencing in 2010 of a US geologist, Xue Feng, to eight years in prison under China's powerful state secrets laws for "stealing secret information about the Chinese oil industry" (Hook 2010a). The secret information that Dr Xue was convicted of stealing was actually a publicly available oil industry database on 30,000 Chinese oil wells belonging to CNPC. He attempted to purchase this information in 2001 when working for IHS Energy, an American consulting firm. This database was classified as a state secret several years after the incident had occurred. From the perspective of international business norms this sort of information on state-run firms is typically regarded as commercial data that should be made publicly available for the purpose of normal market research, and to accord with requirements for listing on international stock exchanges. In China oil industry data is now classified and strictly controlled, safeguarded, restricted and manipulated by Beijing. According to Areddy (2010), analysts have cautioned, "In [China's] oil industry...state-sanctioned publications inexplicably pruned certain data from long-standing charts. Chinese energy company executives increasingly directed industry analysts to channel questions through official spokesmen." They further conclude that it is "best to avoid researching anything about Chinese oil reserves" (Areddy 2010).

The Chinese government's desire to maintain its control over the oil industry has implications for corporate governance. To outside observers the organisational structure of China's NOCs appears similar to that of western firms, complete with clear corporate organisation and boards of directors that include independent directors. However, in practice they operate differently due to the presence of undisclosed CCP instruments of control, including external mechanisms such as the COD, a CCP body that determines personnel appointments, promotions and dismissals in state firms and regulatory bodies, and internal mechanisms such as party committees and powerful party secretaries who often double as company chairmen and CEOs (Yeo 2009b: 1022). These institutions influence personnel management, corporate decision-making and corporate transparency (McNally 2002: 93). Furthermore, the central government's dual role as owner and regulator, which is enacted through state agencies such as SASAC, leads to further interference in the corporate governance of SOEs, undermines managerial autonomy and also reduces incentives to regulate effectively (Ewing 2005: 323).

While the separation of government and enterprise management has been nominally established, in practice interference from government agencies and party institutions continue to affect business operations. This is also the case for listed oil companies since the government is the majority shareholder in all cases, and the interests of minority shareholders are not effectively protected (Tomasic and Andrews 2007). 'Corporate governance with Chinese characteristics' is best conceptualised as a product of gradualist economic reform, whereby western-style corporate governance and laws have been transplanted to the Chinese context of business and politics. Suffice to say these institutions do not possess sufficient authority to act independently of the party-state, especially when it comes to the strategic sectors of the Chinese economy. Hence China's NOCs are driven by both political and commercial agendas. At the same time as the NOCs are expected to meet various national objectives in terms of energy, economic development, and foreign and security policy, they are also responsible to their investors and shareholders in the publicly listed subsidiaries, and are subject to international regulations (Yergin 2011: 206). Yergin (2011: 206) claims that as a result the NOCs are hybrids, "...somewhere between the familiar 'international oil companies,' IOCs, and the state-owned national oil companies, NOCs. They have become a prime example of a new category called INOCs – the international national oil companies."

Ownership and regulation of China's NOCs

While China's NOCs largely possess operational autonomy defined as day-to-day management, the government maintains strategic control of these companies and is able to influence their behaviour through various mechanisms, a particularly significant one being ownership rights. That being said, this chapter does not advance the simple assertion that ownership automatically translates into control. Rather instances are presented that support the argument that ownership rights in combination with other significant mechanisms, notably the nomenklatura system, financial support from state-owned banks, and so on, enables the Chinese government to compel the NOCs to undertake particular activities that further national over commercial interests. Despite the fact that China's NOCs are now listed on both domestic and international stock exchanges, they remain, in essence, government-controlled entities. The key mechanism that ensures government control is share ownership, as with this dominance via majority share ownership comes the capacity to determine the composition of the board of directors and the management of the company (the next section of this chapter discusses the distinctive nature of senior personnel appointments within these firms, which are determined by the COD of the CCP). While the drive to improve operational efficiency through limited exposure to market forces was one of the goals behind the partial privatisation that occurred when the NOCs' IPOs were launched from 2000–2001, the party-state certainly did not intend to relinquish control over the strategic direction of the NOCs and their listed subsidiaries. China's government chose to retain majority ownership of the NOCs due to their strategic significance, and although internationally listed, PetroChina, Sinopec and CNOOC are still treated by the government as state firms since the dominant shareholders are their parent companies (see Table 7.1).

The parent or holding companies are owned and regulated by central SASAC. As explained in the previous chapter SASAC was created in 2003 to unify the state's ownership representation. These Chinese state-owned companies are not homogenous, and consist of large SOEs in strategic sectors under the direct control of central SASAC (currently 117 state firms), the thousands of subsidiaries of these 117 corporations, companies owned by provincial and municipal governments, and companies that have been partially privatised, yet retain the government as the major shareholder (SASAC website). China's NOCs fall under the purview of central SASAC, and hence are subject to more

Table 7.1 Major share ownership of Sinopec, PetroChina and CNOOC

Company	Major Shareholder	Percentage of Shares Owned	Share Type
China Petroleum and Chemical Corporation (Sinopec Ltd)	China Petrochemical Corporation (Sinopec Group)	70.84	State-owned
PetroChina Company Ltd	China National Petroleum Corporation (CNPC Group)	86.07	State-owned
CNOOC Ltd	China National Offshore Oil Corporation (CNOOC)	64.41	State-owned

Source: Table derived from Jia and Tomasic (2010), with updated figures from official company websites.

direct forms of central government supervision and control. The central party-state's oversight of the holding companies through SASAC reduces corporate transparency and makes the task of independent audit by investment advisors very difficult to achieve, as McNally (2002: 109) claims, "Information regarding the performance of state holding corporations is viewed as highly sensitive, because it reflects how the state sector as a whole performs." SASAC does not focus exclusively on exercising the state's ownership, management and supervisory functions, but also assumes substantial regulatory responsibilities that are significant in terms of their impact on the business operations of SOEs (Naughton 2006: 15). One example of SASAC's use of regulatory authority was its decision in April 2005 to forbid management buyouts in China's SOEs (*The Economist* 2005c). Hence SASAC's mandate differs from that prescribed in the *OECD Code for Governance of State Owned Enterprises* (2005), which advises against combining ownership and regulatory functions, in order to establish regulatory independence (OECD 2009: 59). SASAC plays a major role in restructuring industrial organisation and determining the nature of domestic competition among state firms, deliberately building some competition into the state-owned sectors of the Chinese economy (OECD 2009: 59). The domestic market within which the NOCs operate is an oligopoly, where the three major firms – CNPC, Sinopec and CNOOC – partake in limited competition. This oligopolistic structure was established in the 1998 oil industry reform, for which the impetus was to create a limited form of competition in the domestic oil market in order to improve efficiency and prepare the NOCs for competition on the international stage, while at the same time maintain the sector's profitability. This domestic market structure provides China's NOCs with two distinctive features. First, they do not rely on exposure to market discipline to improve performances. Rather they respond mainly to top-down policies and reform agendas that initiate corporate reorganisation. Second, their big profits are largely achieved through monopolisation advantages instead of strategies to succeed in market competition.

Share ownership and international listings obviously determine the type of influence to which listed Chinese companies are exposed. For instance, PetroChina's foreign listings do render the firm subject to outside influence at a basic level due to the regulatory framework of international exchanges. However, the degree of outside influence can be limited, for instance, PetroChina is exempt from the listing requirements of the New York Stock Exchange (NYSE) because its parent company CNPC controls more than 50 per cent of the company's share capital, and hence voting power. There are some examples where foreign investors have influenced these companies' investment decisions and operations. Lavelle (2008: 138) claims that according to the provisions of public offerings in Hong Kong and the United States minority investors holding H-shares are required to vote on affiliated transactions. Under these circumstances PetroChina was obliged to seek the approval of its minority shareholders in order to establish a joint venture with its parent firm (Lavelle 2008: 138). In doing so, PetroChina excluded controversial assets in Sudan, Burma, Turkmenistan, Iraq and Syria to ensure minority investor approval would be granted (Lavelle 2008: 138). Such examples are infrequent, as the Chinese government has deliberately kept state ownership of the NOCs at a high level in order to ensure its control, and minimise external influence through the share ownership structure.

China's NOCs are obligated to protect and advance the national interest in certain ways, even when fulfilling this role compromises company profits and efficiency. The Chinese government clearly stipulates the social responsibilities that SOEs must fulfil, which are explicitly stated in *The Management Law of State-Owned Assets* (Liu 2009: 548). For the NOCs the most basic social responsibility they must fulfil is to guarantee a stable oil supply. In some cases this requirement has had a detrimental impact on financial performances. For example, in 2008

PetroChina faced losses of around US$18 billion on its refining businesses since the costs of soaring international oil prices could not be passed onto consumers because Beijing capped fuel prices to limit inflation. However, due to its strong specialisation in upstream oil activities, which offset to some extent the downstream losses, only Sinopec received subsidies in 2008 to the tune of US$7.6 billion. Jiang Jiemin, chairman of PetroChina at the time, stated that his company was forced to bear the burden of refining losses due to social responsibilities: "We... have to shoulder some social responsibility, such as ensuring the supply of refined oil products in China. So when there is some conflict, some sacrifice will be required" (Crooks 2008). From April 2008 both PetroChina and Sinopec were paid monthly state subsidies against their monthly refining losses (*China Economic Review* 2008). These monthly subsidies replaced the *ad hoc* subsidy payments, which began in 2005, in order to help these companies better manage their cash flows and investment plans. However, compensation provided by the government has not completely recovered refining losses both Sinopec and PetroChina sustained over the years when international oil prices peaked. Mounting refining losses in China's NOCs caused some angst among international shareholders, and contributed to Warren Buffet's (Berkshire Hathaway) decision to liquidate his US$2.3 billion shareholding in PetroChina in 2007 (*ChinaStakes* 2008).

The issue of oil refining output during the oil price hikes that occurred from 2003 to 2008 reveals a divergence of national and corporate interest. Sinopec in particular came into conflict with the party leadership when its chairman Chen Tonghai made the decision to reduce refining output in order to mitigate the company's refining losses. Sinopec's actions resulted in fuel shortages, thus threatening social stability and angering the party leadership. It seems to be no coincidence that less than a year after these events occurred Chen became the subject of a corruption inquiry and resigned from his position at Sinopec. Chen was subsequently expelled from the CCP and handed a suspended death sentence for taking more than US$28 million in bribes. According to McGregor (2010: 64), industry insiders have little doubt that Chen's insubordination regarding refining output in 2005 was the primary cause of his downfall. McGregor (2010: 64) states when "Chen took on senior leaders during the fuel crisis, he handed his enemies an excuse to bring him down". Apparently Chen's corrupt practices were well known and had been going on for some time. The nature of Sinopec's party-appointed replacement, Su Shulin, lends further credence to this view. Su's experience in the oil industry was mainly in party bodies rather than business operations. He was essentially sent to re-install party discipline and control, which the authorities felt Sinopec had lost under Chen (McGregor 2010: 64). Under Su's tenure Sinopec has developed a reputation among analysts and investors for being less transparent than CNPC and CNOOC (Hook 2011c).

It is not uncommon in China for those corruption charges laid against prominent businessmen and government officials to reflect power struggles involving the party leadership. In other words these corruption charges are often a disguise for political purges (MacDonald 2009). For instance, the new administration under Xi Jinping has launched an assault on China's top oil executives, dubbed the 'petro purge', as part of its broader drive to take on those powerful vested interests that hinder reform efforts in China's oil industry, and elsewhere in the state sector. Xi has focused the party-state's efforts on reducing the NOCs' power base and their resistance to further state sector reform, and also tackling endemic corruption (among other political and economic imperatives). Four senior CNPC executives came under investigation for corruption, as well as CCP Central Committee member Jiang Jiemin (former CNPC chairman and director of SASAC), the highest-ranking official to face corruption charges to date (Li 2013; Zhang and Ng 2013; Holland 2013; Downs 2013; Page et al. 2013). While it is highly likely that this is indeed a 'petro purge' designed to eliminate political opposition and consolidate Xi's power and authority, a significant motivation is also to remove powerful state sector figures that have been resistant to further SOE reform, particularly reform that is oriented towards weakening the state sector monopolies. Furthermore, the Chinese leadership is aware that corruption among party officials is beginning to erode the legitimacy of the regime and hinder economic development, and must be dealt with effectively at all levels of government. Hence these measures used to reconfigure power within the party-state to suit the objectives of the top Chinese leadership will probably also have an impact on the structure of the oil industry if it enables deeper market-oriented reforms to proceed.

Another example of the state's control of the NOCs, and use of SOEs to pursue non-commercial objectives, is the construction of the West-East Pipeline Project (WEPP), which delivers natural gas from Xinjiang to Shanghai. In addition to the desire to exploit natural gas reserves in Xinjiang, this pipeline was a part of Beijing's 'Go West' campaign, which aimed to develop China's impoverished western interior, hence fulfilling a major social policy objective (*BBC* 2001). These energy security and social policy considerations rather than commercial imperatives of the NOCs drove the project. Construction of the US$5.2 billion pipeline began in 2002, and was opposed by CNPC's subsidiary PetroChina due to doubts concerning the project's economic viability and Premier Zhu Rongji's insistence on foreign involvement. Despite these commercial objections the Chinese government forced CNPC to undertake the WEPP. Downs (2008a: 131) claims, "Under pressure from Zhu, the company reluctantly issued a tender for foreign participation and signed nonbinding agreements with Shell, ExxonMobil, and Gazprom". Ultimately foreign investors withdrew from the project in large part due to Beijing's insistence on capping the price of gas for domestic industrial users such as large petrochemical companies (Hoyos et al. 2004). While PetroChina welcomed the withdrawal of foreign involvement this development was not a result of lobbying on the part of PetroChina. Construction of the pipeline went ahead without foreign involvement, and the knowledge that any shortfall in meeting costs would be shouldered by the Chinese government (Hoyos et al. 2004). Again this speaks to the centre's capacity to occasionally force SOEs to undertake unprofitable projects.

NOC leadership

The distinctive nature of corporate governance practices in China's NOCs is reflected in company leadership. Executives in China's NOCs tend to be drawn from government bureaucracy and the party. Party secretaries continue to play a leading role in SOEs, also doubling as company chairman. Both McNally (2002: 105) and Chan (2009: 49–50) claim that the membership of the party committee, management team and board of directors is essentially fused in state holding corporations. The party core group forms the centre of decision-making power in the holding companies, and its members hold top positions in management and on the board. The predominant influence of party institutions gives primacy to adhering to "national economic guidelines" and policy directives, over "seeking innovative solutions to increase efficiency" (McNally 2002: 109). The personnel appointment system for SOEs is administered through COD, which remains one of the most powerful and secretive institutions in China (McGregor 2010: 71). In addition to control, the nomenklatura system is the "main source of systemic coherence that strengthens central authority by creating incentives for party members to adhere to central edicts" (Yeo 2009b:

1021). Moreover, party core groups are established in SOEs to further ensure the application of enterprise nomenklatura as a control mechanism. Chan (2009: 5) claims, "Party units in all SOEs are given full authority to make almost all fundamental decisions related to management, personnel, key projects and finance". While SASAC is nominally responsible for personnel appointments, promotions and dismissals, in practice the COD is the "ultimate nominator", which may take advice from SASAC, since the latter has access to better information on suitable candidates (Yeo 2009b: 1021). While party loyalty is a requirement for career advancement in SOEs, the Chinese leadership is determined to cultivate professionally competent SOE management. This has been a key challenge given the lack of domestic market competition among large SOEs. SASAC is responsible for motivating the managers of SOEs in the absence of market-based managerial incentives. Its primary tool for providing effective managerial incentives has been through its three-year performance contracts (Naughton 2010: 453). These contracts outline annual and three-year targets, and performances are graded. Managerial salaries are tied to the grade, and the overall performances of firms are published – an incentive structure that works reasonably well (Naughton 2010: 453).

The boards of directors of the listed subsidiaries are generally composed of executive and non-executive directors who are also executives or senior managers in the parent companies or drawn from their connected companies (see Table 7.2 for the board composition for PetroChina, Sinopec and CNOOC). Jia and Tomasic (2010: 68) suggest that this pattern indicates, "the management of a listed company was very much intertwined with the management of its mother company". A smaller number of directors on the boards of the listed companies are nominally independent, but the extent of independence is difficult to assess. Furthermore, the same person usually occupies the company positions of chairman and CEO. For instance, Fu Chengyu acted as chairman of CNOOC Group, and both chairman and CEO of CNOOC Ltd up until 2010, when he resigned from the role of CEO. In addition Fu was also Party Secretary, reinforcing the CCP's influence on the company. The government selects board members of the listed oil companies, since it is the majority shareholder. Some of the board members often hold various senior management roles within the listed company as well. The chairman of the board of the listed company is also usually the chairman of the board of the parent company, again indicating that there is little practical separation between the two enti-

Table 7.2 Composition of the Board of Directors in China's NOCs

Company	Board Members	Executive Directors	Non-Executive Directors	Independent Directors
Sinopec Ltd	15	10	0*	5*
PetroChina Ltd	14	3	6	5
CNOOC Ltd	11	3	3	5

Source: Official company websites
* Sinopec non-executive directors are classified as independent directors by the company, despite the government and party affiliations of each

ties. While independent directors constitute 30–45 per cent of the boards of the NOCs, they are typically drawn from the government and party elite, except in the case of CNOOC Ltd. By their association with the major shareholder these directors are not independent according to classic corporate governance definitions. For instance, Sinopec's independent non-executive directors all worked for the government and are party members. This leads Jia and Tomasic (2010: 115) to conclude that "Superficially, board composition is an indication of the company's new governance structure; however, a close look at the board's composition suggests that the old management team of the state-owned enterprise remains in place; this might be referred to as 'filling a new bottle with old wine'."

CNOOC is widely considered to be the most "outward-looking, nimble and entrepreneurial" of China's NOCs, and one of China's best managed firms (*Knowledge@Wharton* 2012; *The Economist* 2005a). It is also perceived to be the most commercially-oriented Chinese oil company, appearing to adhere to stronger corporate governance, as Herberg states, "CNPC and Sinopec are more like government ministries [in the process of] becoming like companies," while "CNOOC learned from the very beginning [how to operate internationally]..." (quoted in *Knowledge@Wharton* 2012). As a western-trained manager with extensive experience dealing with western firms, the company's former chairman and CEO Fu Chengyu has played an important role in the company's success. Fu espouses free market rhetoric, and has been quoted as saying "Transparency makes shareholders love you" (*The Economist* 2005a). The incorporation of four foreign independent directors, Evert Henkes, formerly of Royal Dutch/Shell, Kenneth Courtis, vice-chairman of Goldman Sachs in Asia, Erwin

Schurtenberger, a former Swiss ambassador to China, and Chiu Sung Hong, an Australian lawyer, further increased the company's credibility. However, the CNOOC-Unocal debacle of 2005 raised questions concerning whether CNOOC is "really a commercially driven firm, with corporate governance able to protect all its shareholders from Chinese political pressure" (*The Economist* 2005a). Fu pursued Unocal as pulling off such a major deal "would bring huge political influence and secure his future" (*The Economist* 2005a). CNOOC's senior management did not inform the board, let alone seek their approval, about the proposed bid for Unocal until a very late stage, leaving the board feeling pressured to rush the deal through due to time constraints. This led to accusations of "questionable corporate governance and scant regard for minority shareholders" (Guerrera et al. 2005). Schurtenberger resigned soon after the board was finally informed, for health reasons. The remaining independent directors hired their own team of advisors, led by the British investment bank N. M. Rothschild, to investigate whether the bid for Unocal was in the interest of all shareholders. Courtis abstained from approving the US$18.5 billion bid to avoid a conflict of interest. While the board, with the exception of Courtis, eventually granted approval, the deal was thwarted by United States Congress, citing among other things, national security concerns, leading CNOOC to withdraw its bid. This case shows that boards are kept out of the loop when it comes to strategically significant issues, and perhaps also that top executives such as Fu may be prepared to flout corporate governance norms and engage in commercially high risk behaviour in order to enhance their chances of currying favour in Beijing and perhaps political promotion.

Evidence of party-state interference in the business operations of Chinese NOCs can be found in the way in which the COD tends to treat the heads of SOEs as "apparatchiks" who can be "shifted around at will" (McGregor 2010: 84). In recent years the Chinese government has reshuffled senior management in the telecommunications, airline and banking industries. In April 2011 a major reshuffle of oil industry leaders occurred, when a sudden announcement was made that top executives of the holding companies would swap jobs with their rivals. On 3 April Chinese state media reported that Su Shulin, chairman of Sinopec Group would depart to take up a prestigious government position as Fujian province's party head, might further indicate the lack of separation between government and enterprise in China. This was followed by a COD announcement on 8 April that Fu Chengyu, chairman of CNOOC Corp, had replaced Su at Sinopec, and Wang Yilin, vice-president of CNPC Corp, had filled Fu's seat at CNOOC. These reshuffles reveal that the government "views managers as much as state bureaucrats as business executives," all of whom serve the same master (Dickie 2007). The government's aim is to discourage head-to-head competition among state firms, prevent the accumulation of too much power among the top executives and remind the top executives of these companies that the party-state has ultimate control. While these job swaps occurred at the parent company level, conflicts of interest arose as the top oil executives kept their old positions at the listed subsidiaries (Yam 2011). For example, Fu became the head of Sinopec, but remained chairman of CNOOC Ltd, and Wang was still director of PetroChina after he became chairman of CNOOC Group. There was also a substantial delay until the listed subsidiaries announced leadership changes in the parent companies. Once the conflict of interest became publicly recognised the relevant executives resigned from their positions in the listed companies. As these changes played out, the independent directors on the boards of CNOOC, Sinopec and PetroChina remained silent. With reference to the independent directors of these companies, the independent director of one listed, Chinese state-owned bank, claimed, "There is a good chance that the independent directors [didn't] have a clue... They probably read about the change on the same day as you and me" (Yam 2011).

The rotation system of top executives and the business-government cadre swap does indeed "raise issues of career incentives" (Gore 2011: ii). Since many of these executives are evidently focused on furthering their political careers and advancing their business careers through party loyalty, then the interests of the enterprise itself may not always be prioritised. Certainly it would seem to provide compelling incentives for state policy objectives to be incorporated into business operations (Gore 2011: ii). That being said, political promotion for top executives of SOEs in China is largely tied to firm performance. Cao et al. (2009) argue that state control and political incentives for career advancement among top executives may in fact indirectly align their interests with the company's commercial interests, further claiming, "That the likelihood that a CEO receives a political promotion exhibits a strong positive relationship with firm performance indicating that political career concerns are consistent with incentives for value maximisation in Chinese SOEs." However, incentives for political promotion may interfere with incentives to "maximise firm value" on occasions when Beijing expressly privileges non-commercial concerns such as energy security, industrial competitiveness and social stability, which occurred in the case of Sinopec's refining losses. Pursuit of political promotion may sometimes encourage top executives to undermine corporate governance, which was shown in the case of Fu's handling of CNOOC's bid for Unocal. While such instances of interference in the NOCs' business strategy and operations occur infrequently, they reveal party-state influence and control of the NOCs. Now that the government-NOC relationship has been evaluated, the following discusses the internationalisation of these companies and their progress in spearheading China's quest to secure foreign oil supply.

The 'go global' policy

By the mid-1990s China's NOCs were essentially spearheading the 'go global' policy, which was adopted as official government policy in 2001. A range of oil challenges at home, combined with external incentives and a permissive international context prompted the NOCs, following the example of CNPC, to 'go international'. Since CNPC inherited all the administrative functions from the disbanded MPI in 1988, it also assumed the role of policy initiator and was charged with crafting China's onshore petroleum policies. It soon became apparent that reliance on domestic oil reserves and the domestic oil market alone was no longer sufficient. Hence CNPC developed a 'transnational operational strategy' in 1991, which involved looking abroad for oil assets (Kong 2010: 41). From 2001 onwards this strategy received official endorsement from the central party-state, which further intensified the overseas acquisitions undertaken by the NOCs. China's NOCs pursue an equity-based oil acquisition strategy, particularly favouring long-term supply contracts. With reference to China's oil investments in Africa, Taylor (2006: 942) notes, "Because China's oil companies are state-owned China is able to pursue this course even if it means outbidding competitors in major contracts awarded by African governments and paying over the odds." This is often referred as a "neomercantilist" approach to oil security, tapping into fears that China is locking up oil supply. While most of this equity oil is currently sold on international markets, the Chinese leadership believes that China can redirect these oil resources home in the event of a major oil crisis or supply disruption (as unlikely as this may be). Chinese NOCs have often been willing to pay a premium to acquire equity oil, which further indicates a statist rather than liberal market approach to energy security. As of 2011 Chinese NOCs were operating in thirty-one countries, with equity production occurring in twenty of them (Jiang and Sinton 2011: 17).

A major outward foreign direct investment (OFDI) push began in 2000 with the 'go global' policy, which encourages Chinese firms to invest abroad with government approval. This policy was officially initiated by former Premier Zhu Rongji (though as we know it was pioneered by CNPC) in his 2000 report to the NPC on the work of the government, and became formal policy through its incorporation in the Tenth Five Year Plan (2001–2005). The Tenth Five Year Plan listed overseas investment by Chinese enterprises as one of the four key thrusts to enable the Chinese economy to 'adjust to the globalisation trend' (OECD 2008b: 83). During this plan the number of companies approved for overseas investment grew at an annual rate of 33 per cent (Xu 2011). Since it was first promulgated the 'go global' policy has been the subject of continual refinement and adjustment by the Chinese leadership. It was further reinforced in March 2006 in a key policy speech to the annual plenum of the Chinese People's Political Consultative Conference, in which Wen claimed that the government would "institute a policy support and service system and improve the mechanisms for coordinating overseas investment and risk management" (Lunding 2006: 5). A key priority has been to make sure that China's external activities, in particular its OFDI, continue to serve the country's domestic interests.

This was the theme of the Central Work Conference on Foreign Affairs (FAWC) in August 2006, which addressed some of the negative consequences or fallout of the 'go global' policy, with an emphasis on improving central coordination of China's external activities with domestic priorities (Glaser 2006). At this conference one Chinese official commented, "There have been many negative reactions to Chinese foreign policy around the world" (Xu 2011). Of particular concern has been the operational behaviour of Chinese firms in developing countries, including issues such as dumping of Chinese goods and other unfair trade practices, poor treatment of local workers by Chinese companies and harmful environmental impacts. As another Chinese official maintained, "Chinese businesses are going out into the world and they lack knowledge about the world. They have demonstrated bad behaviour. They ignore the local conditions. People have criticised their behaviour as representative of the Chinese government's behaviour" (Xu 2011). This perceived need to improve the 'go global' policy was echoed by Hu Jintao at the Seventeenth Party Congress in 2007, and reflects the party-state's capacity for self-evaluation and ability to learn and adapt to new conditions accompanying the global activities of Chinese firms.

In 2011 SASAC promulgated new rules to further regulate outbound investment by SOEs under its administration, amid 'increasing complexities' (*China Daily* 2011b). While these regulations are aimed at protecting the security of overseas assets it is indicative of Beijing's desire to support, protect and encourage outbound investment to fuel its economic growth and increasing demand for energy and natural resources. The 'go global' policy also features prominently in the Eleventh and Twelfth Five Year Plans (2006–2010) and (2011–2015) respectively, and has become a part of the annual economic development plans passed by the NPC (Devonshire-Ellis 2010). During the time of the Twelfth Five Year Plan the 'go global' policy became more high profile due to the rapid increase and expansion of major overseas investment deals conducted by China's large SOEs. The overseas investments undertaken by China's NOCs in particular became a major focus of international attention. Chinese OFDI increased by 1,400 per cent from 2003 to 2009, and by the end of 2009 China's total non-financial FDI stood at more than US$240 billion (Xu 2011). Hence while western OFDI continues to shrink in the post-GFC era, China's is expanding at a staggering rate. This trend is set to continue, with experts predicting that China's OFDI will soar to US$1 trillion by 2020 (Xu 2011). To date most Chinese FDI has been concentrated in Asia, Latin America and Africa. However, the "new phase" of the 'go global' policy sees increasing Chinese FDI directed toward the United States and Europe (Xu 2011).

While the NOCs are the main executors of the 'go global' policy, the CDB and China Eximbank also play important roles by providing cheap financing for energy and natural resource projects. Both banks are wholly state-owned and charged with advancing China's national interests, which includes supporting the 'go global' policy (Downs 2011: 2). Together these banks signed loans of around US$110 billion to other developing country governments and companies in 2009 and 2010 (Dyer and Anderlini 2011). In comparison the World Bank made loan commitments of US$100.3 billion from mid-2008 to mid-2010 (Dyer and Anderlini 2011). The CNPC alone has been the recipient of an undisclosed loan worth US$30 billion in 2009 at a discounted interest rate from the CDB to fund its overseas upstream oil acquisitions (Downs 2011: 27). In addition to purchasing oil through equity, many of these agreements include 'loans-for-oil' deals with the NOCs and government entities in Russia, Venezuela, Brazil and Kazakhstan (Anderlini 2011). Hence Beijing has been able to promote and enhance the commercial interests of its NOCs by offering loans to oil producing countries (Anderlini 2011). China now has more than US$3 trillion in foreign exchange reserves, and is eager to convert this into hard assets that serve the country's economic development goals. These loans-for-oil deals equip Beijing and the NOCs with a powerful tool to obtain foreign oil assets from cash-strapped oil-producing countries (Jiang 2009a: 9).

Another notable loans-for-oil deal was concluded between China's former vice president Xi Jinping and President Hugo Chavez in 2009, whereby a Chinese-backed bilateral development fund would grow from US$6 billion to US$12 billion, giving Venezuela access to hard currency and the welcome opportunity to diversify oil sales away from the United States (Jiang 2009a: 9). The CDB contributed two-thirds of the capital, and Venezuela the remaining one-third. In return Venezuela guarantees oil shipments to China of one million barrels per day, up from a level of around 380,000 barrels (Romero and Barrionuevo 2009). Even proponents of the FA model such as Downs concede that the CDB's loans-for-oil or 'energy-backed loans' (EBLs) indicate a 'fairly high degree' of coordination between government and business (Downs 2011: 58). Downs (2011: 58) notes that the CDB, China's NOCs and senior government officials work together "to negotiate the various agreements that comprise the deals with their foreign counterparts". The CDB and China Eximbank are policy banks that were established in 1994 and are responsible for financing economic and trade development and state-invested projects. Chen Yuan, the chairman of the CDB, is almost treated like a head of state when he travels abroad, and it is increasingly evident that CDB loans provide Beijing with financial leverage over borrowers to pursue its energy security interests. Certainly China's NOCs have emerged as "a major force in global mergers and acquisitions over the past five years as they carry out their mandate to improve China's energy security by pinning down new supplies of oil" (Hook 2011a).

Beijing's oil diplomacy

Chinese oil diplomacy has, over the past decade, become much more sophisticated, multidimensional, coordinated and flexible, and this has largely occurred in order to facilitate and support the 'go global' policy. Some analysts have even concluded that oil diplomacy now forms the cornerstone of China's foreign policy (Xu 2011). While China's diplomatic efforts have been primarily directed toward improving the country's energy and resource security, there are certain underlying principles governing China's approach to international relations that ought to be mentioned. Key to understanding the success of China's recent diplomatic efforts towards oil-rich countries in the developing world, and in the African and Central Asian regions in particular, has been its 'no political strings attached' policy of engagement, in combination with the financial incentives for cooperation that it offers. This policy is consistent with China's long-running foreign policy traditions of non-intervention and non-interference in the domestic affairs of sovereign states. Hu Jintao updated these foreign policy positions reformulating them into the 'harmonious world' concept (*hexie shijie*) based on 'democracy among nations'. These foreign policy principles are obviously attractive to autocrats the world over, many of whom preside over oil-based economies. China also emphasises, especially in the case of its dealings with Africa, the fact that it shares a common identity as a developing country that has also been a victim of European imperialism. Taylor (2006: 939) argues that China deliberately promotes the suspicion that many developing countries harbour towards western-centric norms of governance; "China taps into this suspicion, asserting that human rights such as 'economic rights' and 'rights of subsistence' are the main priority of developing nations and take precedence over personal, individual rights as conceptualised in the West." Hence there is a sharp divergence between western public diplomacy, which continues to emphasise "rogue states" and "arcs of extremism", and the Chinese position of non-interference and reluctance to criticise other governments (Taylor 2006: 946).

China has also pursued a diplomatic strategy of utilising broader multilateral frameworks for cooperation in targeted regions in order to further safeguard its energy interests. Multilateral cooperation is a key instrument of soft power, which is used effectively by Beijing to facilitate trade and investment in general, as well as bilateral oil deals in regions such as Central Asia and Africa. The Shanghai Cooperation Organisation (SCO), which is composed of China, Russia, Kazakhstan, Kyrgyzstan, Tajikistan and Uzbekistan, is a notable example. Created in 2001 with the SCO Secretariat located in Beijing, this organisation has been actively promoted by China to help achieve its energy objectives in Central Asia. Under this multilateral umbrella China completed a major oil pipeline linking Kazakhstan and Xinjiang (Deng 2008: 219). In more recent years SCO members have sought greater energy cooperation, and have agreed to establish a unified energy market for oil and gas exports, and promote regional growth through preferential trade agreements. While most substantive Central Asian energy cooperation occurs outside of the SCO framework on a bilateral basis, broader multilateral cooperation and aid programs increase China's influence in the region through the economic gains it can bring to Central Asian countries. As Kurlantzick (2007: 199) suggests, China can use the SCO as leverage to present itself as a "natural leader of the region", and it appears to be working. Kazakhstan is a particularly significant supplier of equity oil for China, signing more than a dozen oil production deals with Chinese companies and receiving loans worth US$10 billion pledged against future energy supplies (Gorst 2011). Most recently Kazakhstan signed a currency swap deal with China, agreeing to finance US$1 billion worth of trade in renminbi and tenge by 2015 (Gorst 2011).

The Forum on China-Africa Cooperation (FOCAC) is another significant example of regional multilateral cooperation that is largely driven by China's hunt for energy and natural resources abroad. China has sought to increase dramatically its oil imports from the continent as part of its diversification strategy; 85 per cent of Africa's exports to China come from five oil-rich countries; Angola, Equatorial Guinea, Nigeria, Congo and Sudan (Hanson 2008). China aims to establish special relationships with African countries in the hope that this will result in preferential access to oil resources for China's NOCs. Taylor (2009: 938) points out that China's diplomacy in Africa is spearheaded by the drive to acquire oil, which is underwritten by two main concerns: (1) the short-term goal of securing oil supplies to China, and (2) the desire to position China as a "global player in the international oil market". African oil, along with Central Asian and Latin American oil, is part of China's diversification strategy to reduce dependency on oil from the Middle East. China's 'Africa strategy' is multifaceted and is "illustrative of Beijing's efforts to create a paradigm of globalisation that favours China" (Brautigam and Tang 2009: 143). It is strikingly clear in the case of FOCAC how Beijing integrates diplomacy, aid and business. Since it first took place in 2000 in Beijing, FOCAC has been held every three years, and according to China's 'African Policy', "has become an effective mechanism for the collective dialogue and multilateral cooperation between China and Africa" (FOCAC website). FOCAC also works with African countries to implement the *Beijing Declaration of the FOCAC, Program for China-Africa Cooperation in Economic and Social Development* and the *Forum on China-Africa Cooperation – Addis Ababa Action Plan (2004–2006)*. The contents of these documents suggests that China-Africa cooperation is intended to expand at various levels and into fields including economic, trade, financial, agricultural, medical care and public health, transportation, environment and tourism. According to the Beijing Declaration China-Africa cooperation will establish "within the framework of South-South cooperation a new-type long-term and stable partnership based on equality and mutual benefit" (FOCAC website). The FOCAC Summit held in Beijing in 2006 was a watershed event in Sino-African relations. Forty-eight out of fifty-three African countries attended the elaborate forum, held in 'China's Year of Africa' and in commemoration of the fiftieth anniversary of China's diplomatic relationship with Africa (Jiang 2008: 19).

This summit saw China pledge extensive financial support and assistance to underscore future China-Africa cooperation such as the provision of US$3 billion in preferential loans over the course of three years, the cancellation of US$1 billion of debt from African countries, the establishment of special economic zones in Africa and a China-Africa Development Fund amounting to US$5 billion, the signing of trade deals worth US$1.9 billion, a grant of US$37.5 million for anti-malarial drugs, and assistance in building thirty hospitals and one hundred rural schools (FOCAC website; Jiang 2008: 19–20). China has made good on these pledges, and has embarked upon further concessional loan and debt cancellation initiatives in Africa since then. China's oil diplomacy is evidently pursued at the highest political levels, with Hu Jintao actively promoting Chinese trade and investment with Africa in a series of high profile tours beginning in 2006. For instance, in 2007 Hu visited eight African countries – Cameroon, Liberia, Sudan, Zambia, Namibia, South Africa, Mozambique and Seychelles – on a twelve-day tour, during which he offered aid and debt cancellations, and signed numerous bilateral cooperation agreements (*China Daily* 2007; Jiang 2008: 19). The combination of financial incentives multilateralism and China's 'no strings attached' policy has indeed enhanced China's presence and influence on the African continent, resulting in booming trade and investment over the past decade – China-Africa trade has grown from a mere US$6 billion on 1999 to just over US$100 billion in 2010 (Rice 2011). Hence it is evident that China's multipronged diplomatic efforts have augmented the country's OFDI regime, and has helped Chinese NOCs secure long-term oil supply contracts abroad.

This chapter argued that the relationship between the central party-state and the NOCs is characterised by collaboration governed by hierarchy. This conceptualisation of the relationship is not intended to imply that the Chinese government micromanages these firms, nor controls their day-to-day activities and decision-making. Chapter 5 examined how the government largely divested itself of these responsibilities during the 1980s and 1990s. The NOCs possess operational autonomy and Beijing has sought to transform them into commercially-oriented and internationally competitive enterprises, with reasonable success. Nevertheless, the party leadership maintains key levers of control, which it can wield in order to influence the strategic direction of the NOCs, and command certain decisions to be taken where it serves the national interest. In order to ascertain the degree of control over the NOCs that Beijing possesses, it is best to examine those instances where there has been a divergence of commercial and national/party-state interests. Most of the time there is a convergence of commercial and national interest on the part of China's NOCs and the central party-state – generally the NOCs going about their business in a commercially-oriented manner also serves Beijing's interests. However, there are compelling counter-examples where there has been a clear divergence, the most striking being Sinopec's decision to reduce its oil refining output in order to stem losses that were increasing as international crude oil prices soared while domestic end-user prices were capped at a much lower level. In the end Sinopec was forced to increase output and its chairman was punished for his actions, which the Chinese leadership perceived as an insubordinate act that endangered social stability. The Chinese leadership also pursues a strategy of using political purges to remove rival political factions and obstacles to economic development. The 2013 'petro purge' is partly motivated by the Xi administration's desire to push ahead with state sector reforms that are opposed by key figures now under investigation for corruption. Naughton (2010: 457) neatly summarises the situation for political and business leaders in the state sector, "Officials and managers have lots of authority and strong temptations, but they are subject to strong surveillance and draconian punishments." SASAC also regularly swaps and reshuffles the top executives of the large SOEs including the NOCs, indicating that ultimately these chairmen and CEOs serve Beijing and the CCP.

These cases, and the others presented in this chapter, demonstrate that the central party-state retains significant authority and control over the NOCs, and can shape their strategic direction through a variety of mechanisms. In addition, the political and financial support extended to the NOCs, which is further augmented by Beijing's oil diplomacy, provides the NOCs with a significant risk subsidy and competitive edge over rival oil companies in the international arena. With regard to the central role of the NOCs in improving China's energy security, Pei (2006a) states that the NOCs "now have practically unlimited financial resources to make overseas acquisitions, regardless of economic viability or geopolitical risks. If their deals go bad, Beijing will foot the bill. The companies will not be blamed, either, because they are simply implementing the government's energy security strategy. But if they strike 'black gold', the companies reap the economic benefits." Hence the integration of energy security objectives with foreign policy, combined with almost 'unlimited' financial support, has certainly enhanced the capacity of these companies to engage in global competition.

8
Authoritarian State Capacity in a Liberal World Order

The main objectives of this study were to provide an explanation of why China pursues a state-led oil strategy, and show how this orientation has been developed and implemented over time. This involved an examination of China's oil state capacity-building efforts with a focus on the Reform Era, and including a discussion of the Mao era of oil industry development, since it left a legacy of useful policy instruments and bureaucratic capacity, which then formed the foundation of post-Mao industrial development. A sophisticated range of political, organisational and fiscal capacities enables Beijing to pursue a statist oil agenda, and these have been strengthened and expanded during the second Reform Era in particular. At this stage in China's overall development and transition to a market-oriented economy, state-led oil policies are considered a necessary strategy to safeguard economic growth, industrial competitiveness, and social harmony and stability. Repeating the words of Kreft (2006), quoted earlier in this volume, the Chinese leadership believes that the oil sector is strategically "too important to be left to market forces alone". The central party-state has demonstrated a core interest in improving the performance of the oil industry through the introduction of market characteristics and institutions, while at the same time maintaining control over this vitally important strategic sector. The party leadership's attempts to improve China's oil state capacity have produced reasonably effective and more comprehensive oil policies. This view differs from conventional scholarly accounts (see, for example, Downs (2004a and 2006); Kong (2006 and 2010); Meidan et al. (2009); Lester and Steinfeld 2006 and 2007), which present a dismal view of energy policymaking and oil sector governance in China. They argue that the country's allegedly fragmented and decentralised bureaucracies produce suboptimal oil policy outcomes. Such a view is informed by the FA model, which of course suggests that the structure of the Chinese state and political system places major constraints upon policymakers. According to proponents of this explanatory model, the fragmented and decentralised structure of Chinese bureaucracy tends to produce policy inertia and can also distort policy at the implementation phase. Within this institutional context powerful vested interests, such as the NOCs, are said to wield disproportionate influence upon the policy process. Hence scholars such as Downs (2008a) and Houser (2008) claim that China's domestic and international oil policy behaviour is the product of bottom- or middle-up initiatives, rather than top-down political authority. From this perspective China's oil policy approach is not really state-led, but rather is led by the NOCs.

This volume offers a counter-argument to the conventional understanding of energy governance in China as it pertains to the oil industry, which relies on an alternative conceptualisation of the Chinese policy process. It has argued that during the second Reform Era China successfully strengthened and expanded various political and organisational capacities to improve the coherence, coordination and effectiveness of its energy policies. Institutional arrangements in China, both at the industry-level and nationwide, have not remained unchanged across the decades of reform, as is implied by FA (this is essentially a static model that cannot readily account for institutional change). In particular China's institutions have been reshaped and reorganised to suit the exigencies of market transition. In the case of the commanding heights of the Chinese economy, institutions have been adjusted to accommodate the party-state's desire to both marketise these sectors and also retain centralised control over them. The allegedly constraining effects of institutions have been largely overcome in the party leadership's drive to restrengthen and expand the capacity of the 'centre' and also rebuild capacity in the strategic sectors. In the case study chapters it has been demonstrated that the country's political elites have the capability to reach in and restructure the Chinese state and its institutions, thus providing an agent-centred and elitist account of institutional change. The theoretical framework that best describes this policy process is BA. This model recognises that China's political system is primarily characterised by top-down, hierarchical and centralising principles of governance. During the second Reform Era decision-making at the top became much more decisive and China's political elites sought to reconstitute the hierarchy of the central party-state, which had dissipated somewhat during the first Reform Era. The Tiananmen protests and the collapse of communism in the Soviet Union and Eastern Bloc prompted widespread capacity-building efforts in various parts of the party-state in order to consolidate the power and ruling capacity of the CCP. The following decade saw major rounds of bureaucratic restructuring and the establishment of new institutions of economic governance to govern an increasingly market-oriented economy as well as advance a more state-led economic agenda (such as in the case of oil).

The extent of institutional change during the second Reform Era provides a clear indication that the party leadership can indeed overcome bureaucratic resistance where sufficient political will exists to push through their policy agendas. In the case of the oil sector, such political will did not really begin to form until around 2003 when energy security shot to the top of Beijing's policy agenda following energy crises and a host of new energy challenges. Many of the major changes in oil sector organisation can be mapped onto wider transformations in China's economy, including the drive to establish a modern enterprise system, leading to the creation and eventual internationalisation of China's NOCs during the 1980s and 1990s. In most cases such changes can be traced to party-state elites remaking the institutions that govern the oil industry and the economy more broadly, to meet the imperatives not only of energy security but, more fundamentally, economic growth and regime survival. Another relevant feature of the Chinese policy process is the leadership's preference to pursue a gradualist or incremental reform agenda. This gradualist reform agenda is neither a product of bureaucratic opposition nor indicative of a 'weak centre' as the FA model suggests, but rather of a political culture that places a premium upon the maintenance of stability, harmony and order. Gradualist reform allows the central party-state to iron-out problems along the way, and gives bureaucratic actors time to learn and internalise new rules and norms in order to ensure smoother transitions. Hence institutional change in China tends to occur incrementally, although the party leadership can readily switch to a more decisive mode of policymaking when "changes in external conditions demand a rapid response" (Naughton 2008a: 104). In terms of oil policymaking a combination of stability and decisiveness has been evident, where the Chinese leadership has taken active and direct involvement in devising China's energy policies and restructuring the oil industry to improve performances and create internationally competitive NOCs, albeit gradually across the decades of reform.

A significant portion of the literature that deals with China's energy security is devoted to examining the relationship between the Chinese government and the NOCs. Here Downs (2008a) and Houser (2008) argue that the 'tail wags the dog', that is, the NOCs shape China's energy policies and may not always act in accordance with Beijing's interests. Kong (2010) claims that the NOCs and the Chinese government co-govern China's oil industry, which suggests a relationship characterised by equality. These views of the government-NOC relationship are again very much informed by the FA model, and ignore the power of the central party-state to solicit compliance from the NOCs through a variety of political, organisational and financial mechanisms. In contrast, this study defined the relationship as collaboration governed by hierarchy, where the steep hierarchy of the central party-state structures the interaction between the government and the NOCs. According to this conceptualisation the flow of political authority is primarily top-down, however, the NOCs are able to advise the party leadership on oil strategy and have influenced policy content and implementation. Throughout the first decade of the twenty-first century the central party-state has gradually built institutional capacity by streamlining and centralising administrative functions, creating institutions to support market-oriented development and structure domestic competition among the NOCs, improving coordination among various energy-related bureaucracies, and producing clearer and more coherent energy policies at the highest levels of government (espoused in the Tenth, Eleventh and Twelfth Five Year Plans, the 2007 energy white paper, and a variety of other policy documents). Correspondingly, China's state-led oil policies have indeed become more comprehensive, sophisticated, coordinated and integrated with other energy goals, such as those pertaining to efficiency and conservation, over time. Importantly, China's oil policy approach is also used by the party-state to advance other social and economic objectives, particularly with regard to social equity and industrial competitiveness. This largely accounts for why the party leadership wishes to retain control of this lifeline industry, as it impacts almost every other sector of the Chinese economy, is used to manage inflation (through the suppression of energy prices), and, as a result, is intimately linked to the continuing survival of the CCP as the country's ruling body. This is not to say that FA has ceased to function as a relevant model of Chinese policymaking, but that its effects can be mitigated effectively by features of the central party-state that it fails to factor into consideration.

The rise of China's market authoritarianism model

At a deeper conceptual level this study also comments on the strength of authoritarian state capacity, and the ability of illiberal political regimes to survive, if not flourish, within a globalised liberal world order. Throughout Asia, Russia, Latin America and the Middle East governments are harnessing state capitalist models, often in combination with tight autocratic control, to pursue economic development. The Chinese model is the prime example of an authoritarian political system that has attempted to wed capitalism with a large and interventionist role for the state in the economy. The rise of authoritarian state-directed capitalism goes against the consensus that had previously been formed in the West, which claimed that liberal democratic transition is inevitable once a certain level of economic and social development has been achieved. This teleology of modernity, famously championed after the end of the Cold War by Fukuyama (1992), has gained currency since the third wave of global liberal democratic expansion and continues to influence western studies on contemporary Chinese governance. A large body of literature continues to assume that China's political future will ultimately be a democratic one, and if democratic transition does not occur then the country will fail to achieve its development aspirations and either implode under the weight of a corrupt and dysfunctional political system or lapse into stagnation (see, for instance, Pei 2006b). In addition, authoritarian political systems have in the past been typically viewed as inherently fragile and unstable due to their "weak legitimacy, over-reliance on coercion, overcentralisation of decision-making, and the predominance of personal power over institutional norms" (Nathan 2003: 6). It is now clear that such teleological speculation about the 'end of history' was premature, a point that was driven home in an interview with Kagan (2008), where he said:

> We lived under the illusion that economic success required political liberalisation. All the optimism of the 1990s rested on this assumption. Now it appears that the causality is less certain. Autocratic governments can sustain economic growth, and indeed their economic success helps them sustain their autocracy. This means, if nothing else, that we must be ready for a world in which powerful autocracies endure and perhaps even thrive.

In addition Friedman (2000: 226) claims that "Democratisation is a matter of grasping infrequent opportunities," noting that such opportunities in recent history have actually presented themselves under conditions of economic decline or implosion, notably in the cases of Latin America and Leninist command economies in the Eastern Bloc and former Soviet Union. Far from converging around western liberal democracy and free market capitalism, the political order remains diverse, and while levels of interdependence among states are high, genuine integration and convergence has not been forthcoming. In terms of China's political future, Nathan (2003) was one of the first China scholars to identify China's 'authoritarian resilience', which seemed to indicate that China was successfully adapting to modernity without triggering a transition to democracy.

These conventional assumptions have often resulted in somewhat blinkered research when it comes to the issue of political reform in China. The predominant Western perception is that there has been an absence of political reform, and the Chinese political system remains a stagnant and 'ossified Leninist state' (Shambaugh 2008: 2). Shambaugh (2008: 2–3) considers this inaccurate view to be based upon the widespread scholarly presumption that if reforms are not democratically-oriented, they are not valid and are automatically discounted as ineffectual. However, this view simply does not bear scrutiny upon closer examination of China's record during the second phase of reform, which shows that the CCP has been proactive in instituting illiberal, yet often very effective, capacity-building reforms within itself and the Chinese state (Shambaugh 2008: 2–3). The observation that illiberal adaptation and authoritarian resilience can strengthen within the current international political context reveals a basic failing of the democratisation literature in particular, which typically deems democratisation to be a historically inevitable and universal process. In light of China's rise and a wave of 'democratic retrenchment' or 'backsliding' in the developing world (Diamond 2008), it would seem that liberal democracy may not be the only political system that is compatible with modernity. Furthermore, the foundations of the western liberal order, that is, free markets and democratic pluralism, have been damaged over the past decade. Halper (2010: 34–35) claims that the actions of Coalition nations in Iraq and Afghanistan have "stained both the moral authority of the West and the notion that democratic pluralism is universally viable". The GFC in 2008 and continuing economic woes in the United States, Britain and the European Union, have led to a crisis of confidence in neoliberalism and, more generally, in western ideas and institutions of economic and financial management (Halper 2010: 34–35). These developments have precipitated a decline in the West's, and especially America's 'soft power', that is, the ability to lead by example and shape the preferences of others through attraction and persuasion, rather than coercion. Arguably the GFC also highlighted the need for strong and effective state institutions to oversee financial markets.

Hence China's state capitalist model presents a challenge to western ideas on politics and economics, and many other developing countries are now shifting in a more authoritarian direction. While stopping well short of claiming that the Chinese model can be exported to countries in regions such as Latin America or Africa, Halper (2010: 32) argues that China is "exporting something simpler, and indeed more corrosive to Western pre-eminence... This is the basic idea of market authoritarianism." Wolf (2008) makes the interesting observation that a very large proportion of the wealth currently being accumulated by economies is ending up in state coffers. Paradoxically, the 'great liberation of global capitalism' has produced more powerful state owners than ever before. This seems especially peculiar in light of the effort expended by many western governments and Washington-based financial institutions in attempting to eliminate public ownership. Related to this, the international political and economic implications of Chinese state-owned firms operating within a capitalist economy are increasingly the subject of debate. State-directed economies and their SOEs have functioned effectively within a liberal international economic order before with examples that include South Korea's state-led development strategy and Singapore's state-controlled firms.

However, the case of China is different; it is now the world's second largest economy, and its SOEs, such as the NOCs are behemoths that "until now have been inward-looking but are starting to use their vast resources abroad" (*The Economist* 2010a). China's 'go global' policy, which encourages SOEs to go abroad and secure energy and natural resources, rather than compete openly for them in the free market, has picked up pace since it was first promulgated in 2001. China has the fiscal capacity to pursue such a policy on a grand scale, particularly as it needs to invest its enormous trade surplus and savings abroad. At the same time traditional sources of foreign investment – the United States, Britain and Japan – are continually shrinking. This has resulted in a situation where the world's fastest growing new sources of wealth and investments are non-western, and often non-democratic. While often erroneously compared with Japan, in terms of rising levels of ODI, China is qualitatively different, as it is a country ruled by an opaque authoritarian government, which is commonly regarded as a strategic competitor of the West. A column in

The Economist (2010a) neatly summarises the concerns surrounding China's OFDI:

> Resources would be allocated by officials, not the market. Politics, not profit, might drive decisions. Such concerns are being voiced with increased fervour. Australia and Canada, once open markets for takeovers, are creating hurdles for China's state-backed firms, particularly in natural resources, and it is easy to see other countries becoming less welcoming too.

This study revealed that the oil sector in China is subject to a higher degree of central party-state control and management than it was even a decade ago, which is the result of a broader drive to reinforce central political authority within the Chinese state, and particularly within the strategic sectors. Whether this level of political control poses a fundamental threat to the liberal capitalist system is difficult to predict in the long term. That being said, China's quest to secure foreign oil assets, which is now occurring on a large scale, provides some indication as to what the impact on markets and governance might be at the international level.

Implications of China's state-led oil strategies for business and politics

The extant literature that deals with the implications China's international energy behaviour can be broadly divided into two main streams; the China energy threat, and the business as usual view. The China threat theory, as it pertains to energy security, tends to focus on geostrategic energy issues such as maritime brinkmanship over oil resources between China and other states in disputed waters in the South China Sea. Another subset looks at China's mercantilist approach of "locking up" energy resources, which implies heightened competition, and even conflict, over scarce oil resources. These sorts of concerns revolve around the potential for China's activities to exacerbate competition among oil-consuming states, and NOCs and IOCs, with the possibility of conflict occurring. However, in practical terms such conflicts can, and have been managed, although as East Asia transitions to a more multipolar regional order as the result of declining US hegemonic power, interstate relations are likely to be more volatile and unstable. Hence flashpoints will continue to occur, but the chance of such tensions escalating to major conflict among the great powers is slim. The other prominent view in the literature argues that China's energy behaviour, and the activities of Chinese NOCs in particular, will have little impact on the international economy, both in terms of the functioning of markets and governance practices. Again this view tends to assume that as China embraces capitalism it will also choose to conform to western rules, norms and institutions. Proponents of this view uncritically assume that Beijing has chosen to 'integrate' into the extant US-led international order rather than challenge it in any meaningful way, and as such advocate greater engagement with China.

In light of the research undertaken for this study it is reasonable to suggest that the challenge from China's state capitalist model in general, and state-led oil strategies in particular, is much more complex and, perhaps less obvious, than conventional arguments concerning the China threat suggest. However, the argument presented by the optimists that China's rise would have little impact on the current international order based on western rules, norms and institutions, is also rather uncritical and simplistic. The view that China will ultimately integrate and accept these extant market and governance arrangements is not a foregone conclusion, and there is much evidence to the contrary. With regard to the issue of Chinese integration within a US-led international system, Mann (2007: 105) asks the important question, "'Who's integrating whom?' Is the United States now integrating China into a new international economic order based upon free market principles? Or, on the other hand, is China now integrating the United States into a new international political order where democracy is no longer favoured and where a government's continued eradication of all organised political opposition is accepted or ignored?". China is predicted to become the world's largest economy, surpassing America's, within the next decade, and if it retains its model of market authoritarianism, then foreign governments and businesses may need to change to accommodate China. For example, private companies such as Yahoo and Microsoft have altered their own rules and operations by complying with the Chinese Web-censorship regime in order to do business in China (Mann 2007: 105). In a rare show of defiance, Google took a principled stand against Chinese censorship, and pulled its Internet search engine out of China in 2010 (*The New York Times* 2010: A26). It is also becoming evident that some of China's business practices and 'no political strings attached' policy potentially undermine western-led institutions of global economic governance such as the IMF, World Bank and WTO. When China perceives its national interest is better served by ignoring the rules established by these institutions, then it will flout them. For example, the United States Trade Representative's *2010 Report to Congress on China's WTO*

Compliance claims, "significant questions have arisen regarding China's adherence to ongoing WTO obligations, including core WTO principles. Frequently, these problems can be traced to China's pursuit of industrial policies that rely on excessive, trade-distorting government intervention intended to promote or protect China's domestic industries and state-owned enterprises."

China is increasingly turning to other developing countries for trade and investment, which further erodes the relevance of western influence, particularly since many of these developing economies are also pursuing development models that reject economic liberalism and democratic institutions. Countries in Africa increasingly view Chinese investment and aid as preferable to that provided by western countries and economic institutions. This is because China's pursues a 'no political strings attached' investment strategy, in contrast to the requirements concerning human rights standards, good governance and transparency that tend to accompany western investment and aid. Hence China's growing presence in Africa, driven by its need for energy and natural resources, is said to undermine liberal principles of business and politics in this region, and erode the influence of institutions such as the World Bank and IMF. That being said, the failure of these western-led institutions to generate economic development and growth in the developing world over the past thirty years has produced widespread disillusionment with western aid projects, and more enthusiastic acceptance of new sources of aid and investment, especially from China. Booming Chinese trade with Africa, which hit US$114.81 billion in 2010, has given African economies a much-needed boost (MOFCOM 2011). China's commitment to financing the construction of critical infrastructure (ports, roads and so on) in Africa has also strengthened in recent years – a development that should attract further FDI to countries in this region (FOCAC 2010).

While China's growing presence in Africa may hinder western efforts to promote liberal norms, it is transforming African economies. In this pursuit of energy and natural resources abroad China has shown that it is prepared to deviate from global norms regarding the type of regimes it chooses to deal with. For example, China's oil dealings with Sudan from 2007–2011 attracted international criticism, since during that time it contravened western sanctions on the Sudanese government that were meant to put pressure on Khartoum for its handling of the crisis in the Darfur region. More recently China's opposition to a US-led oil embargo against Iran has invited criticism from Washington. This embargo was designed to put pressure on Tehran to abandon a suspected nuclear weapons program. Iran is China's third largest supplier of crude oil (roughly 500,000 barrels per day) hence losing Iran's oil imports would cause an immediate supply shock to China (Pei 2012). China's commercial activities in these countries arguably reduce the effectiveness of these sanctions. Kong (2010: 149), along with other energy security scholars who claim China's oil sector is characterised by weak state capacity and the problems associated with bureaucratic fragmentation, attempt to make the argument that China does not intentionally flout global norms with regard to markets and governance, even in cases such as Sudan. Here the argument is that China's NOCs are independent from the Chinese government and will enter foreign countries, such as Sudan, even against the will of the party-state in order to pursue their commercial objectives. While it is difficult to disprove such a view, from a BA perspective the central party-state would have certainly possessed the capacity to step in and force the NOCs to withdraw from Sudan if that was what they really desired. Furthermore, a very large proportion of China's equity oil now comes from Sudan and South Sudan, an outcome that is just as much the product of Chinese government policy as commercial activity on the part of the NOCs (Shichor 2008: 73). Certainly the argument that Beijing cannot control the activities of NOCs can be used as a convenient defence to ward off criticism of China's approach to pariah states.

It would be fair to say that China's NOCs have altered the global oil industry's competitive landscape, where IOCs are arguably now placed at a relative disadvantage. In order to fuel its commodity-intensive growth engine China actively encourages its NOCs to secure oil resources in Asia, Latin America and Africa and is prepared to provide extensive financial and political support for oil equity investments abroad. As a result China's NOCs often out-bid their rivals, sometimes quite spectacularly, in order to secure oil supply for China, and are provided with a risk subsidy. While most of this equity oil is then sold on the world market, rather than shipped directly to China, the important point for the Chinese leadership is to own oil in the ground so that it can be redirected to China in the event of an oil crisis (as unlikely as this may be). China's NOCs also possess the ability through this strong state support to formulate long-term strategies in line with the national interest. In contrast the commercial or economic objectives that drive IOCs place them at a disadvantage *vis-à-vis* NOCs, since they must concentrate on short-run profitability, and cannot take the sorts of risks that NOCs are prepared to undertake (since ultimately they are shouldered and absorbed by the Chinese state). Hence Beijing essentially provides Chinese NOCs with a risk subsidy that is simply unavailable to IOCs. In addition, major resource holders are, more often than not, developing countries with autocratic political systems that increasingly find Chinese investment more attractive than western investment, for reasons already stated. Beijing's oil diplomacy and broader aid and investment packages offered to host countries also sweeten the oil deals that are pursued by Chinese NOCs.

In sum, Beijing's efforts to build oil state capacity during the second Reform Era have resulted in the development and reasonably successful implementation of more coordinated and effective oil policies (and energy policies more broadly). This shift is generally not accounted for in the extant literature, which continues to adhere to the FA model. While the FA model offers significant insights on the Chinese policy process, its narrow focus on mid-level bureaucratic structures misses the powerful political tools that are used by Beijing to solicit compliance from bureaucracies within the party-state. The Chinese leadership shows a clear preference for state-led oil strategies, considering them appropriate to safeguard China's economic growth and social stability at this particular stage of the country's development. This study is more broadly situated within emerging debates over a major shift in the global economy that has occurred due to the rise of state capitalist economies such as the BRICS (Brazil, Russia, India, China and South Africa), and concomitant decline of liberal market economies, notably the United States and Britain. China's growing investments in developing countries arguably diminish the influence that western countries can wield in order to promote norms relating to good governance and human rights. Furthermore, the fabric of global capitalism is changing, with growth increasingly dependent on state-led or public investment, which is often accompanied by protectionist or neomercantilist impulses. Halper (2010: 8) neatly summarises this phenomenon: "The economic pre-eminence of the West is being increasingly moderated by the new norms and networks of the rest; the 'invisible hand' of free markets around the globe is being balanced by the notably more visible hand of central governance." These trends fundamentally alter the competitive landscape within which western or private sector businesses operate, especially when it involves direct competition with SOEs.

At an even deeper level China's ascendancy challenges the assumed universality of western values, rules and institutions, giving rise to an era of what Jacques (2009: 144) refers to as 'contested modernity', whereby new paths to modernity will not be automatically discounted as backward or ineffective compared to the western model. Rather than providing the 'unquestioned model' for the rest of the world to follow, the West might offer just one of several possibilities. Jacques (2009: 144–145) argues that:

> In the future [the West] will be required to think of itself in relative rather than absolute terms, obliged to learn about, and to learn from, the rest of the world without the presumption of underlying superiority, the belief that ultimately it knows best and is the fount of civilisational wisdom... The emergence of Chinese modernity immediately de-centres and relativises the position of the West.

This volume has shown that in the case of China's oil sector, the central party-state has pursued a statist oil strategy from the Mao era through to the present day. During the post-Mao or Reform Era oil industry reform has centred on introducing market characteristics to improve efficiency and performances, while at the same time retaining central party-state control. Since roughly 2003 energy security has commanded a great deal of attention from the Chinese leadership and this led to renewed efforts to build oil state capacity, thereby improving oil policymaking and implementation with a focus on guiding, promoting and supporting the international activities of the NOCs. The capacity of the central party-state to adapt a successful state capitalist model runs contrary to conventional (neoliberal) assumptions regarding the efficacy, or lack thereof, of strong state intervention in the economy. If the CCP continues to succeed in sustaining economic growth, while at the same time improving governance and the provision of public goods, then demand for democratically-oriented political reforms within China will remain low. When couched within broader debates concerning the rise of state-directed capitalism, this argument also contributes to the claim that state-directed capitalism arguably provides a credible alternative to the neoliberal model.

Bibliography

Aizhu, Chen and Jim Bai. "China's Power Woes Give Little Impetus to Oil Prices." *Reuters*, January 12, 2010.
Alden, Chris and Christopher R. Hughes. "Harmony and Discord in China's Africa Strategy: Some Implications for Foreign Policy." In *China and Africa: Emerging Patterns in Globalization and Development*, edited by Julia C. Strauss and Martha Saavedra. Cambridge: Cambridge University Press, 2009.
Alexander, Ruth. "Which is the World's Biggest Employer?" *BBC News*, March 20, 2012.
Allison, Tony. "Investors Courting China's CNOOC." *Asia Times Online*, November 10, 2000: http://www.atimes.com/reports/BK10Ai01.html
Anderlini, Jamil. "On Good Terms: Chinese Banks Fuel 'Going Global' Drive." *The Financial Times*, April 5, 2011.
—— "China's State Sector Urged to Prop Up Cooling Domestic Economy." *The Financial Times*, December 27, 2008.
Andrews-Speed, Phillip. *The Governance of Energy in China: Transition to a Low-Carbon Economy*. Basingstoke: Palgrave Macmillan, 2012.
—— *China, Oil and Global Politics*. Hoboken: Taylor & Francis, 2011.
—— *The Institutions of Energy Governance in China*. Gouvernance européenne et géopolitique de l'énergie. IFRI, January 2010.
—— "China's Drive for Energy Efficiency." *Far Eastern Economic Review* 172, no.3 (2009): 33–38.
—— *Energy Policy and Regulation in the People's Republic of China*. The Hague: Kluwer Law International, 2004a.
—— "State Control is the Cause of China's Crisis." *Asian Wall Street Journal*, April 30, 2004b.
Andrews-Speed, Stephen Dow and Zhiguo Gao. "A Provisional Evaluation of the 1998 Reforms to China's Government and State Sector: The Case of the Energy Industry." *Journal of the Centre for Energy, Petroleum and Mineral Law and Policy* 4, no.7 (1999): 1–11.
Andrews-Speed, Liao Xuanli and Roland Dannreuther. *The Strategic Implications of China's Energy Needs*. Adelphi Paper No. 346. New York: International Institute for Strategic Studies, 2002.
Areddy, James T. "China's Culture of Secrecy Brands Research as Spying." *The Wall Street Journal*, December 1, 2010.
Arruda, Michael E. and Ka Yin Li. "China Energy Sector Survey Part III: Foreign Inbound Investment." *China Law & Practice* (February 2004): 19–27.
—— "China Energy Sector Survey Part II: The Energy Institutions." *China Law & Practice* (December 2003/January 2004): 19–28.
—— "China's Energy Sector: Development, Structure and Future." *China Law & Practice* (November 2003): 12–17.
Baeg Im, Hyug. "The Rise of Bureaucratic Authoritarianism in South Korea." *World Politics* 39, no.2 (1987): 231–257.

Baum, Richard. *Burying Mao: Chinese Politics in the Age of Deng Xiaoping.* Princeton: Princeton University Press, 2004.
—— *Reform and Reaction in Post-Mao China: The Road to Tiananmen.* New York: Routledge, 1991.
BBC, "China's Premier Chairs Energy Meeting 20 April." *BBC Monitoring Asia,* April 22, 2006.
—— "Premier Zhu Says Cross-Country Pipeline Symbol of China's Opening Up." *BBC Monitoring Asia-Pacific – Political,* July 5, 2001.
Beeson, Mark. "Politics and Markets in East Asia: Is the Developmental State Compatible with Globalization." In *Political Economy and the Changing Global Order,* edited by Richard Stubbs and Geoffrey R. D. Underhill. New York: Oxford University Press, 2006.
Bell, Stephen. "Institutionalism: Old and New." In John Summers (ed.) *Government, Power and Policy in Australia* (7th ed.). NSW Australia: Pearson Education Australia, 2002: 363–380.
Bell, Stephen and Andrew Hindmoor. *Rethinking Governance: The Centrality of the State in Modern Society.* New York: Cambridge University Press, 2009.
Bell, Stephen and Hui Feng. *The Rise of the People's Bank of China: The Politics of Institutional Change.* Cambridge, MA: Harvard University Press, 2013.
—— and Hui Feng. "Reforming China's Stock Market: Institutional Change Chinese Style." *Political Studies* 57 (2009): 117–140.
—— and Hui Feng. "Made in China: IT Infrastructure Policy and the Politics of Trade in Post-WTO China." *Review of International Political Economy* 14 (2007): 49–76.
Bernstein, Richard and Ross Munro. *The Coming Conflict with China.* New York: Vintage Books, 1998.
—— "The Coming Conflict with China." *Foreign Affairs* 76, no.2 (1997): 18–32.
Blair, Bruce, Yali Chen and Eric Hagt. "The Oil Weapon: Myth of China's Vulnerability." *China Security* 2, no.3 (2006): 32–63.
Bo, Zhiyue. "China's New National Energy Commission: Policy Implications." *EAI Background Brief No.504,* February 5, 2010.
BP (British Petroleum). *BP Statistical Review of World Energy June 2013.* London: British Petroleum, 2013.
—— *BP Statistical Review of World Energy June 2012.* London: British Petroleum, 2013.
—— *BP Statistical Review of World Energy June 2011.* London: British Petroleum, 2012.
Brautigam, Deborah A. and Tang Xiaoyang. "China's Engagement in African Agriculture: 'Down to the Countryside'." In *China and Africa: Emerging Patterns in Globalization and Development,* edited by Julia C. Strauss and Martha Saavedra. Cambridge: Cambridge University Press, 2009.
Brewer, John. *The Sinews of Power: War, Money and the English States, 1688–1783.* New York: Knopf, 1989.
Broadman, Harry. *Africa's Silk Road: China and India's New Economic Frontier.* Washington D.C.: The World Bank, 2007.
Brødsgaard, Kjeld E. and Zheng Yongnian. *Bringing the Party Back In: How China is Governed.* Singapore: Eastern Universities Press, 2004.
Cai, Peter Yuan. "China's New National Energy Commission." *East Asia Forum,* March 12, 2010.

Calder, Kent. *Asia's Deadly Triangle: How Arms, Energy and Growth Threatens to Destabilize Asia-Pacific*. London: Nicholas Brealey, 1997.
Canning, Cherie. "Pursuit of the Pariah: Iran, Sudan and Myanmar in China's Energy Security Strategy." *Security Challenges* 3, no.1 (2007): 47–63.
Cao, Xiaping J., Michael Lemmon, Xiofei Pan and Gary Tian. "Political Promotion, CEO Compensation, and Their Effect on Firm Performance." *Research Collection Lee Kong Chian School of Business*, Paper 1816 (2009).
Carmody, Padraig R. and Francis Owusu. "Competing Hegemons? Chinese versus American Geo-economic Strategies in Africa." *Political Geography* 26, no.5 (2007): 504–524.
Chan, Hon S. "Politics Over Markets: Integrating State-Owned Enterprises into Chinese Socialist Market." *Public Administration and Development* 29, no.1 (2009): 43–54.
Chandler, Clay. "Can China Keep the Lights On?" *Fortune* 149, no.4 (2004): 116.
Chang, Gordon G. *The Coming Collapse of China*. London: Arrow, 2002.
Chang, Ha-Joon. "The Economic Theory of the Developmental State." In *The Developmental State*, edited by Meredith Woo-Cummings. Ithaca, New York: Cornell University Press, 1999.
Chen, Matthew E. "National Oil Companies and Corporate Citizenship: A Survey of Transnational Policy and Practice." Essay/monograph in the Baker Institute energy study *The Changing Role of National Oil Companies and International Energy Markets* (2007).
Chen, Shaofeng. "Has China's Foreign Energy Quest Enhanced its Energy Security?" *The China Quarterly* 207 (2011): 600–625.
—— "Motivations Behind China's Foreign Oil Quest: A Perspective from the Chinese Government and the Oil Companies." *Journal of Chinese Political Science* 13, no.1 (2008): 79–104.
—— "State-Regulated Marketization: China's Oil Pricing Regime." *Perspectives* 7, no.3 (2006): 151–172.
Chen, Shu-Ching Jean. "Price Controls Again in Vogue Among China's Energy Planners." *Forbes.com*, January 17, 2008.
Cheng, Joseph. "A Chinese View of China's Energy Security." *Journal of Contemporary China* 17, no.55 (2008): 297–317.
China Daily. "Jiang Jiemin Removed from Office." September 3, 2013.
—— "China to Select Third Phase Oil Reserve Location." August 23, 2011a.
—— "China Tightens Supervision on Overseas Assets." June 26, 2011b.
—— "China Still Leading Clean Energy Investment." March 30, 2011c.
—— "Wang Yong Takes Helm at SASAC." August 25, 2010a.
—— "Wen Heads 'Super Ministry' for Energy." January 28, 2010b.
—— "3 Chinese Firms Break into Fortune 500's Top Ten." July 9, 2010c.
—— "China to Set Up Five New Super Ministries." March 11, 2008.
—— "Chinese President Wraps up Africa Tour." February 11, 2007.
—— "China Sets Up National Energy Leading Group." June 4, 2005a.
—— "Leading Group to Oversee Energy Sector." March 7, 2005b.
—— "Government Rules Out Forming New Energy Ministry." December 2, 2004a.
—— "CPC Issues Document on Ruling Capacity." September 27, 2004b.
—— "Growing Demand, Inefficiency Blamed for China's Energy Shortages." August 14, 2004c.

—— "Energy Sector Reform Urged." January 8, 2004d.
—— "Four Factors Cause Power Shortage." December 7, 2003.
China Economic Review. "What's Up with the Corruption Crackdown?" September 5, 2013.
—— "NDRC Will Continue to Set Energy Prices." March 25, 2008.
China-Gov.cn. "President Hu Urges Efforts to Ensure Global Energy Security." July 17, 2006.
ChinaStakes. "PetroChina Share Price a Victim of Speculation, Government Control." September 1, 2008. Available at: http://www.chinastakes.com/2008/11/petrochina-share-price-a-victim-of-speculation-government-control.html
Choi, Hyun Jin. "Fueling Crisis or Cooperation? The Geopolitics of Energy Security in Northeast Asia." *Asian Affairs* 36, no.1 (2009): 3–28.
Christoffersen, Gaye. "The Dilemmas of China's Energy Governance: Recentralization and Regional Cooperation." *The China and Eurasia Forum Quarterly* 3, no.3 (2005): 55–79.
Clover, Charles. "Xi Turns to Russia for First Foreign Trip." *The Financial Times*, March 22, 2013.
Clark, Martin. "Willing to Go Where Western Companies Fear to Tread." *The Financial Times*, January 28, 2008.
Collier, David. "Bureaucratic Authoritarianism." In *The Oxford Companion to the Politics of the World* (2nd edition) edited by Joel Krieger. Oxford: Oxford University Press, 2001: 93–95.
Collier, David, Henry E. Brady and Jason Seawright. "Sources of Leverage in Causal Inference: Toward an Alternative View of Methodology." In *Rethinking Social Inquiry: Diverse Tools, Shared Standards*, edited by David Collier and Henry Brady. Lanham: Rowman and Littlefield, 2004.
Constantin, Christian. "Understanding China's Energy Security." *World Political Science Review* 3, no.3 (2007): 1–29.
—— "China's Conception of Energy Security: Sources and International Impacts." Working Paper No.43 (2005).
Cornelius, Peter and Jonathan Story. "China and Global Energy Markets." *Orbis* 51, no.1 (2007): 5–20.
CPC News. July 11, 2013. Available at: http://cpc.people.com.cn/n/2013/0711/c64094-22164183.html
Crooks, Ed. "PetroChina Pays for Oil's Surge." *The Financial Times*, March 19, 2008.
Dannreuther, Roland. "Asian Security and China's Energy Needs." *International Relations of the Asia-Pacific* 3, no.2 (2003): 197–219.
Dean, Mitchell. *Critical and Effective Histories: Foucault's Methods and Historical Sociology*. London: Routledge, 1994.
Deng, Yong. *China's Struggle for Status: The Realignment of International Relations*. New York: Cambridge University Press, 2008.
Development Research Center of the State Council. *China's National Energy Strategy*. Beijing, November 2003.
Development Research Center of the State Council and the Energy Research Institute. *China National Energy Strategy and Policy 2020*. Beijing, June 2004.
Devonshire-Ellis, Chris. "China's Plenum on Next Five-Year Plan? Actually, More of the Same." *China Briefing*, October 19, 2010.

Diamond, Larry. "The Democratic Rollback." *Foreign Affairs*, March/April 2008.
Dickie, Mure. "Cadre Exchange Stirs Telecom Rumours." *The Financial Times*, July 11, 2007.
Dickson, Bruce J. "Updating the China Model." *The Washington Quarterly* 34.4 (2011): 39–58.
—— "The Future of China's Party-State." *Current History* (September 2007): 243–245.
—— *Red Capitalists in China: The Party, Private Entrepreneurs, and Prospects for Political Change*. Cambridge: Cambridge University Press, 2003.
Dingli, Shen. "China's Energy Problem and Alternative Solutions." *Journal of Contemporary China* 10, no.29 (2001): 717–722.
Dittrick, Patrick. "China's NOCs Go Shopping." *Oil & Gas Journal* 104, no.4 (2006): 15–19.
Dittmer, Lowell. *China's Continuous Revolution*. Berkeley: University of California Press, 1987.
Downs, Erica. 'A Tiger Hunt in China's Oil Patch: Why Are So Many Former Oil Executives Under Investigation?' *Brookings*, September 13, 2013.
—— "Inside China, Inc.: China Development Bank's Cross-Border Energy Deals." John L. Thornton China Center Monograph Series, The Brookings Institution, no.3 (March 2011).
—— "Business Interest Groups in Chinese Politics: The Case of the Oil Companies." In *China's Changing Political Landscape*, edited by Cheng Li. Washington D.C.: Brookings Institution Press, 2008a: 121–141.
—— "China's 'New' Energy Administration." *China Business Review* (November/December 2008b): 42–45.
—— "China's Quest for Overseas Oil." *Far Eastern Economic Review* 170, no.7 (2007): 52–56.
—— "China." *The Brookings Foreign Policy Studies Energy Security Series* (December 2006).
—— "The Chinese Energy Security Debate." *The China Quarterly* 177 (2004a): 21–41.
—— "China's Energy Security." Diss., Princeton University, 2004b.
Dyer, Geoff and Jamil Anderlini. "China's Lending Hits New Heights." *The Financial Times*, January 17, 2011.
Ebel, Robert E. *China's Energy Future*. Washington D.C.: Center for Strategic and International Studies, 2005.
Erickson, Andrew S. and Gabriel B. Collins, "China's Oil Security Pipe Dream: The Reality, and Strategic Consequences, of Seaborne Imports." *Naval War College Review* 63, no.2 (2010): 89–111.
—— "Beijing's Energy Security Strategy: The Significance of a Chinese State-Owned Tanker Fleet." *Orbis* 51, no.4 (Fall 2007): 665–684.
Eurasia Group. *China's Overseas Investments in Oil and Gas Production*. New York: Eurasia Group, 2006.
Evans, Peter B. *Embedded Autonomy: State and Industrial Transformation*. Princeton: Princeton University Press, 1995.
——, Dietrich Reuschemeyer and Theda Skocpol. *Bringing the State Back In*. New York: Cambridge University Press, 1985.
Ewing, Richard D. "Chinese Corporate Governance and Prospects for Reform." *Journal of Contemporary China* 14, no.43 (2005): 317–338.

Fenby, Jonathan. *Modern China: The Fall and Rise of a Great Power, 1850 to the Present*. London: Ecco, 2008.
Feng, Jianhua. "Great Expectations." *Beijing Review*, October 11, 2008.
Fewsmith, Joseph. *China Since Tiananmen*. Cambridge: Cambridge University Press, 2008a.
—— "Staying in Power: What Does the Chinese Communist Party Have to Do?" In *China's Changing Political Landscape*, edited by Cheng Li. Washington D.C.: Brookings Institution Press, 2008b: 212–118.
FOCAC. "China Strengthening Africa's Infrastructure Base." Forum on China-Africa Cooperation, March 19, 2010: http://www.focac.org/eng/jlydh/xzhd/t674046.htm
—— "China's African Policy." Forum on China-Africa Cooperation, Beijing, September 20, 2006: http://www.focac.org/eng/zfgx/dfzc/t481748.htm
—— "Forum on China-Africa Cooperation Beijing Action Plan (2007–2009)." November 16, 2006: http://www.focac.org/eng/ltda/dscbzjhy/DOC32009/t280369.htm
—— "Beijing Declaration of the Forum on China-Africa Cooperation" Forum on China-Africa Cooperation, Beijing, 2000: http://www.focac.org/eng/wjjh/hywj/t157833.htm
Friedberg, Aaron L. ""Going Out": China's Pursuit of Natural Resources and Implications for the PRC's Grand Strategy." *NBR Analysis* 17, no.3 (2006): 5–34.
Friedman, Edward. "Immanuel Kant's Relevance to an Enduring Asia-Pacific Peace." In *What if China Doesn't Democratize? Implications for War and Peace*, edited by Edward Friedman and Barrett L. McCormick. Armonk, New York: M. E. Sharpe, 2000.
Friedman, Lisa. "China Leads Major Countries with $34.6 Billion Invested in Clean Energy." *The New York Times*, March 25, 2010.
Fukuyama, Francis. *State-Building: Governance and World Order in the 21st Century*. Ithaca, New York: Cornell University Press, 2004.
—— *The End of History and the Last Man*. London: Penguin, 1992.
Garbaccio, Richard F. "Price Reform and Structural Change in the Chinese Economy: Policy Simulations Using a CGE Model." *China Economic Review* 6, no.1 (1995): 1–34.
Gelber, Harry G. *The Dragon and Foreign Devils: China and the World, 1100 BC to the Present*. New York: Walker & Company, 2007.
Gertz, Bill. *The China Threat: How the People's Republic Targets America*. Washington D.C.: Regnery Publishing, 2000.
Gilley, Bruce. *China's Democratic Future: The Coming Great Leap to Freedom*. New York: Columbia University Press, 2004.
Gittings, John. "Strikes Convulse China's Oil-Rich Heartlands." *The Guardian*, March 21, 2002.
Glaser, Bonnie. "Ensuring the 'Go Abroad' Policy Continues to Serve China's Domestic Priorities." *China Brief* 7, no.5 (2006).
Goldstein, Andrea E. *The Rise of China and India: What's in It for Africa*. Paris: Development Center of the Organisation for Economic Cooperation and Development, 2006.
Goldstein, Avery. "The Diplomatic Face of a Grand Strategy: A Rising Power's Emerging Choice." *The China Quarterly* 168 (2001): 835–864.

—— *From Bandwagon to Balance of Power Politics*. Stanford: Stanford University Press, 1991.
Goldstein, Carl. "Not So Slick." *Far Eastern Economic Review* 157, no.14 (1994): 66–67.
—— "Energy: Lost Privileges." *Far Eastern Economic Review* 156, no.30 (1993): 50–52.
—— "China's Oil Shock." *Far Eastern Economic Review* 155, no.45 (1992): 52–54.
Goodman, Peter S. "Big Shift in China's Oil Policy." *The Washington Post*, July 13, 2005.
Gore, Lance P. "China Recruits Top SOE Executives into Government: A Different Breed of Politicians?" *EAI Background Brief No. 661* (2011).
Gorst, Isabel. "Kazakhstan and China Agree $1bn Currency Swap." *The Financial Times*, June 13, 2011.
Graham-Harrison, Emma. "China Tries to Buck Trend Towards Declining Oil Production." *The New York Times*, April 21, 2008.
Green, Stephen. "'Two-thirds Privatisation': How China's Listed Companies are – Finally – Privatizing." *Chatham House Briefing Note* (2003).
Guerrera, Francesco, Joe Leahy and James Politi. "Twists and Turns in Log of Treasure Ship's Journey." *The Financial Times*, July 14, 2005.
Guo, Sizhi. *The Business Development of China's National Oil Companies: The Government to Business Relationship in China*. Policy Report, Houston: James A. Baker III Institute for Public Policy Rice University, 2007.
Halper, Stefan. *The Beijing Consensus*. New York: Basic Books, 2010.
Halpern, Nina. "Information Flows and Policy Coordination in the Chinese Bureaucracy." In *Bureaucracy, Politics, and Decision Making in Post-Mao China*, edited by Kenneth Lieberthal and David Lampton. Berkeley: University of California Press, 1992.
Hama, Katsuhiko. "The Daqing Oil Field: A Model in China's Struggle for Rapid Industrialization." *The Developing Economies* 18, no.2 (1980): 180–205.
Hamrin, Carol Lee and Suisheng Zhao. "Core Issues in Understanding the Decision Process." In *Decision-making in Deng's China: Perspectives from Insiders*, edited by Carol Lee Hamrin and Suisheng Zhao. Armonk, New York: M. E. Sharpe, 1995.
Hanson, Stephanie. "Backgrounder: China, Africa, and Oil." *Council on Foreign Relations*, June 6, 2008.
Harris, Stuart. "Global and Regional Orders and the Changing Geopolitics of Energy." *Australian Journal of International Affairs* 64, no.2 (2010): 166–185.
Harvey, David. *A Brief History of Neoliberalism*. Oxford: Oxford University Press, 2005.
Heilman, Sebastian and Elizabeth J. Perry (eds). *Mao's Invisible Hand: The Political Foundations of Adaptive Governance in China*. Cambridge, Mass: Harvard University Asia Center, 2011.
Helmer, John. "Kremlin Decides China Pipeline on New Terms." *Asia Times Online*, March 4, 2003.
Herron, James. "The Great EU-China Oil Swap." *The Wall Street Journal*, January 11, 2012.
Hoffman, Philip and Jean-Laurent Rosenthal. "Political Economy of Warfare and Taxation in Early Modern Europe: Historical Lessons for Economic Development." In *The Frontiers of the New Institutional Economics*, edited by John Drobak and John Nye. San Diego: Academic Press, 1997.

Holbig, Heike. "Ideology After the End of Ideology. China and the Quest for Autocratic Legitimation." *Democratization* 20.1 (2013): 61–81.
—— "Remaking the CCP's Ideology: Determinants, Progress and Limits Under Hu Jintao." *Journal of Current Chinese Affairs* 38, no.3 (2009): 35–61.
Holland, Tom. "China Graft Probe Looks More Like a 'Petro Purge' than an Assault on Sleaze." *South China Morning Post*, September 4, 2013.
—— "China's 'Go Out' Policy Means Trouble." *South China Morning Post*, April 26, 2007.
Holslag, Jonathan. "China's New Mercantilism in Central Africa." *African and Asian Studies* 5, no.2 (2006): 132–168.
Hong, Eunsuk and Laixiang Sun. "Dynamics of Internationalization and Outward Investment: Chinese Corporations' Strategies." *The China Quarterly* 187 (2006): 610–634.
Hongyi, Harry Lai. "China's Oil Diplomacy: Is it a Global Security Threat?" *Third World Quarterly* 28, no.3 (2007): 3–22.
Hook, Leslie. "China's Oil Groups Ready for More Deals." *The Financial Times*, August 28, 2011a.
—— "Doing Business in China: A Tale of Two Oil Spills." *The Financial Times*, August 30, 2011b.
—— "Sinopec Chief Tipped for Political Post." *The Financial Times*, March 22, 2011c.
—— "China Jails US Geologist for 8 Years." *The Financial Times*, July 5, 2010a.
—— "China Denies IEA Claim on Energy Use." *The Financial Times*, July 20, 2010b.
Hook, Leslie and Kathrin Hille. "China Chiefs Split Top Roles in Push for Board Independence." *The Financial Times*, August 19, 2010.
Horton, Christopher. "New Revolution Threatens 'Mandate of Heaven'." *Asia Times Online*, March 14, 2003: http://www.atimes.com/atimes/China/EC14Ad01.html
Houser, Trevor. "The Roots of Chinese Oil Investment Abroad." *Asia Policy* 5 (2008): 141–166.
Hoyos, Carlos, Uchenna Izunda and Richard McGregor. "Pipeline Pullout Embarrasses PetroChina." *The Financial Times*, August 4, 2004.
Huang, Jing. "Institutionalization of Political Succession in China: Progress and Implications." In *China's Changing Political Landscape*, edited by Cheng Li. Washington D.C.: Brookings Institution Press, 2008: 80–97.
Huang, Yasheng. "Rethinking the Beijing Consensus." *Asia Policy* 11 (2011): 1–26.
—— *Capitalism with Chinese Characteristics*. New York: Cambridge University Press, 2008.
Huang, Yanzhong. "The State of China's State Apparatus." *Asian Perspective* 28, no.3 (2004): 32–33.
Huntington, Samuel. *Political Order in Changing Societies*. London: Yale University Press, 2006 [1968].
Husain, Syed Rashid. "IEA Courting Beijing to Restore its Eminence in the Energy World." *Arab News*, April 4, 2010.
IEA. *World Energy Outlook 2012*. Paris: OECD/IEA, 2012.
—— *World Energy Outlook 2011*. Paris: OECD/IEA, 2011.
—— *World Energy Outlook 2010*. Paris: OECD/IEA, 2010.

—— World Energy Outlook 2007: China and India Insights. Paris: OECD/IEA, 2007.
Ikenberry, G. John. "The Irony of State Strength: Comparative Responses to the Oil Shocks of the 1970s." *International Organization* 40, no.1 (1986): 105–137.
Ishida, Hiroyuki. "An Analysis of Energy Strategies in China and India." *The Journal of Energy and Development* 32, no.1 (2006).
Jacques, Martin. *When China Rules the World: The Rise of the Middle Kingdom and the End of the Western World*. London: Allen Lane, 2009.
Jaffe, Amy Myers and Steven W. Lewis. "Beijing's Oil Diplomacy." *Survival* 44, no.1 (2002): 115–134.
Jakobsen, Linda and Zha Daojiong. "China and the Worldwide Search for Oil Security." *Asia-Pacific Review* 13, no.2 (2006): 60–73.
Jayasuriya, Kanishka. "Beyond Institutional Fetishism: From the Developmental to the Regulatory State." *New Political Economy* 10, no.3 (2005): 381–387.
Jewell, Jessica. *The IEA Model of Short-Term Energy Security (MOSES): Primary Energy Sources and Secondary Fuels*. Paris: OECD/IEA, 2011.
Jia, Xinting and Roman Tomasic. *Corporate Governance and Resource Security in China*. New York: Routledge, 2010.
Jiang, Julie and Jonathan Sinton. *Overseas Investments by Chinese National Oil Companies: Assessing the Drivers and Impacts*. Paris: OECD/IEA, 2011.
Jiang, Wenran. "China Makes Strides in Energy 'Go-Out' Strategy." *China Brief* 9, no.15 (2009a).
—— "Fuelling the Dragon: China's Rise and Its Energy and Resources Extraction in Africa." In *China and Africa: Emerging Patterns in Globalization and Development*, edited by Julia C. Strauss and Martha Saavedra. Cambridge: Cambridge University Press, 2009b.
—— "Hu's Safari: China's Emerging Strategic Partnership in Africa." In *China in Africa*, edited by Arthur Waldron. Washington D.C.: The Jamestown Foundation, 2008.
—— "Beijing's 'New Thinking' on Energy Security." *China Brief* 6, no.8 (2006).
Jiang, Wenran and Zining Liu. "Energy Security in China's 12th Five Year Plan." *China Brief* 11, no.11 (2011).
Jianxin, Zhang. "Oil Security Reshapes China's Foreign Policy." Working Paper No.9, Center on China's Transnational Relations, 2005.
Johnson, Chalmers. *MITI and the Japanese Miracle: The Growth of Industrial Policy, 1925–75*. Stanford: Stanford University Press, 1982.
Jonquieres, Guy. "China's Industrial Policy Should Think Small." *The Financial Times*, September 7, 2006.
Kagan, Robert. "Interview with Robert Kagan and Gideon Rachman: Illiberal Capitalism." *The Financial Times*, January 17, 2008.
Kahn, Joseph and Jim Yardley. "As China Roars, Pollution Reaches Deadly Extremes." *The New York Times*, August 26, 2007.
Kalicki, Jan H. and David L. Goldwyn. "The Need to Integrate Energy and Foreign Policy." In *Energy & Security: Toward a New Foreign Policy Strategy*, edited by Jan H. Kalicki and David L. Goldwyn. Washington D.C.: Woodrow Wilson Center Press, 2005.
Kambara, Tatsu and Christopher Howe. *China and the Global Energy Crisis: Development and Prospects for China's Oil and Natural Gas*. Cheltenham: Edward Elgar, 2007.
Kang, David. *China Rising: Peace, Power and Order in East Asia*. New York: Columbia University Press, 2007.

Katzenstein, Peter J. "Conclusion: Domestic Structures and Strategies of Foreign Economic Policy." In Peter J. Katzenstein (ed.) *Between Power and Plenty: Foreign Economic Policies of Advanced Industrial States*. Madison, WI: University of Wisconsin Press, 1978.
Keidel, Albert. "China's Looming Crisis – Inflation Returns." *Carnegie Policy Brief*, no.54 (2007).
Kennedy, Andrew B. "China's New Energy Security Debate." *Survival* 52, no.3 (2010): 137–158.
Kerr, David. "Has China Abandoned Self-Reliance?" *Review of International Political Economy* 14, no.1 (2007): 77–104.
Kjær, Anne Mette and Ole Hersted Hansen. "Conceptualizing State Capacity." *DEMSTAR Research Report No.6*. The Department of Political Science, University of Aarhus, 2002.
Klare, Michael and Daniel Volman. "The African 'Oil Rush' and US National Security." *Third World Quarterly* 27, no.4 (2006): 609–628.
Klinghoffer, Arthur J. "Sino-Soviet Relations and the Politics of Oil." *Asian Survey* 16, no.6 (1976): 340–352.
Knowledge@Wharton. "The Return of CNOOC." October 16, 2012. Available at: http://knowledge.wharton.upenn.edu/article/the-return-of-cnooc-2/
Kong, Bo. *China's International Petroleum Policy*. Santa Barbara, California: Praeger Security International, 2010.
—— "Institutional Insecurity." *China Security* (Summer 2006): 65–89.
—— "An Anatomy of China's Energy Insecurity and its Strategies." *Oil, Gas & Energy Law Intelligence* 6, no.1 (2005): 1–69.
KPMG. "China's 12th Five-Year Plan: Energy." *KPMG China* (April 2011).
Krasner, Stephen D. *Defending the National Interest*. Princeton, New Jersey: Princeton University Press, 1978.
Kreft, Heinrich. "China's Quest for Energy." *Policy Review* 139 (2006). Available at: http://www.hoover.org/publications/policy-review/article/7941
Kroeber, Arthur. "Rising China and the Liberal West." *China Economic Quarterly* 12, no.1 (2008): 29–44.
Kurlantzick, Joshua. *Charm Offensive: How China's Soft Power is Transforming the World*. Carlton, Vic.: Melbourne University Press, 2007.
Kwok, Vivian Wai-yin. "China Caps Energy Prices to Contain Inflation." *Forbes.com*, January 10, 2008.
Kynge, James. *China Shakes the World: The Rise of a Hungry Nation*. London: Weidenfeld & Nicolson, 2006.
Lafargue, François. "China's Presence in Latin America: Strategies, Aims and Limits." *China Perspectives* 68 (2006): 2–11.
—— "The Oil War Game in Africa." *Politique Internationale* 112 (July 2006): 401–421.
Laffont, Jean-Jacques and Claudia Senik-Leygonie. *Price Controls and the Economics of Institutions in China*. Paris: Development Center of the OECD, 1997.
Lam, Willy. "The Rise of the Energy Faction in Chinese Politics." *China Brief* 11, no.7 (2011).
—— *Chinese Politics in the Hu Jintao Era: New Leaders, New Challenges*. New York: London: M. E. Sharpe, 2006.
Lampton, David M. *The Three Faces of Chinese Power: Might, Money, and Minds*. Berkeley: University of California Press, 2008.

—— "Paradigm Lost: The Demise of 'Weak China'." *The National Interest* 81 (2005): 73–80.

—— "China's Foreign and National Security Policymaking Process: Is it Changing, and Does it Matter?" In *The Making of Chinese Foreign and Security Policy in the Reform Era*, edited by David M. Lampton. Stanford: Stanford University Press, 2001.

—— ed. *Policy Implementation in Post-Mao China*. Berkeley: University Press, 1987.

Lau, Laurence J., Yingi Qian and Gerard Roland. "Reform Without Losers: An Interpretation of China's Dual-Track Approach to Transition." *Journal of Political Economy* 108, no.1 (2000): 120–143.

Lavelle, Kathryn C. "The Business of Governments: Nationalism in the Context of Sovereign Wealth Funds and State-Owned Enterprises." *International Affairs* 62, no.1 (2008): 131–147.

Lee, Pak. "China's Quest for Oil Security: Oil (Wars) in the Pipeline?" *The Pacific Review* 18, no.2 (2005): 265–301.

Lelyveld, Michael. "China Frets Over Fuel." *Radio Free Asia*, February 28, 2011.

Lema, Adrian and Kristian Ruby. "Between Fragmented Authoritarianism and Policy Coordination: Creating a Chinese Market for Wind Energy." *Energy Policy* 35, no.7 (2007): 3879–3890.

Leonard, Mark. *What Does China Think?* London: Fourth Estate, 2008.

Lester, Richard and Edward S. Steinfeld. "China's Real Energy Crisis." *Harvard Asia Pacific Review* 9, no.1 (2007): 35–38.

—— "China's Energy Policy: Is Anybody Really Calling the Shots?" *Massachusetts Institute of Technology Working Paper Series* (2006).

Levi, Margaret. *Of Rule and Revenue*. Berkeley: University of California Press, 1989.

Li, Cheng. "The End of the CCP's Resilient Authoritarianism? A Tripartite Assessment of Shifting Power in China." *The China Quarterly* 211 (2012): 595–623.

—— "Assessing China's Political Development." In *China's Changing Political Landscape*, edited by Cheng Li. Washington D.C.: Brookings Institution Press, 2008: 1–24.

Li, Raymond. "Former Oil Company Chief and Protégé of Ex-Security Tsar Zhou Yongkang is Latest Graft Probe Target." *South China Morning Post*, September 3, 2013.

Li, Xiaofei. "State Companies Hold Power Over Chinese Energy Policy." *Oil & Gas Journal* 109, no.6 (2011): 26–30.

Liao, Xuanli. "The Petroleum Factor in Sino-Japanese Relations." *International Relations of the Asia-Pacific* 7, no.1 (2007): 23–46.

Lieberman, Joseph. "China-US Energy Policies: A Choice of Cooperation or Collision – Remarks by Joseph I. Lieberman." (Transcript), Council on Foreign Relations, November 30, 2005.

Lieberthal, Kenneth. *Governing China: From Revolution through Reform*. New York: W. W. Norton, 1995.

—— "Introduction: The 'Fragmented Authoritarianism' Model and its Limitations." In *Bureaucracy, Politics, and Decision Making in Post-Mao China*, edited by Kenneth Lieberthal and David Lampton. Berkeley: University of California Press, 1992.

Lieberthal, Kenneth and David Lampton. *Bureaucracy, Politics, and Decision Making in Post-Mao China*. Berkeley: University of California Press, 1992.

Lieberthal, Kenneth and Michel Oksenberg. *Policy Making In China: Leaders, Structures and Processes*. Princeton: Princeton University Press, 1988.

Lim, Tai-Wei. *Oil in China: From Self-reliance to Internationalization*. Singapore: World Scientific, 2010.

Liou, Chih-shian. "Bureaucratic Politics and Overseas Investment by Chinese State-Owned Oil Companies: Illusory Champions." *Asian Survey* 49, no.4 (2009): 670–690.

Liu, Qiao. "Corporate Governance in China: Current Practices, Economic Effects and Institutional Determinants." *CESifo Economic Studies* 52, no.2 (2006): 415–453.

Liu, Yunhua. "A Comparison of China's State-Owned Enterprises and Their Counterparts in the United States: Performance and Regulatory Policy." *Public Administration Review* 69, special issue (2009): 46–52

Lowi, Miriam R. *Oil Wealth and the Poverty of Politics: Algeria Compared*. Cambridge: Cambridge University Press, 2009.

Lu, Li and Emanuela Todeva. "The Petrochemical Industry in China – Government Regulation and Development Policies." In *Proceedings of the Asia-Pacific Research in Organisation Studies Annual Conference*, December 2000.

Lu, Zhenhua. "Wind Power Growth in China's Deserts Ignored Financial Risks." *The Guardian*, May 14, 2010.

Luft, Gal. "US, China on a Collision Course Over Oil." *Los Angeles Times*, February 2, 2004.

Lunding, Andreas. "Global Champions in Waiting: Perspectives on China's Overseas Direct Investment." *Deutsche Bank Research*, Current Issues: China Special (2006).

Ma, Damien. "China's Search for a New Energy Strategy." *Foreign Affairs*, June 4, 2013. Available at: http://www.forcignaffairs.com/articles/139425/damien-ma/chinas-search-for-a-new-energy-strategy

—— "Introducing China's newest Five-Year Plan." *The Atlantic*, October 28, 2010.

Ma, Laurence J. C. "Oil from the Wells of China." *Geographical Review* 70, no.1 (1980): 99–101.

MacDonald, Mark. "Beijing Court Convicts Ex-Sinopec Chief of Bribery." *The New York Times*, July 15, 2009.

MacFarquhar, Roderick. *Origins of the Cultural Revolution* 1. Cambridge, Massachusetts: Oelgeschlager, Gunn, and Hain, 1981.

McGregor, Richard. *The Party: The Secret World of China's Communist Rulers*. London: Penguin, 2010.

—— "China Moves from Hunter to Gatherer." *The Financial Times*, August 21, 2007.

McNally, Christopher A. *China's Emergent Political Economy: Capitalism in the Dragon's Lair*. New York: Routledge, 2008.

—— "Strange Bedfellows: Communist Party Institutions and New Governance Mechanisms in Chinese State Holding Corporations." *Business and Politics* 4, no.1 (2002): 91–115.

Mahoney, James. "Qualitative Methodology and Comparative Politics." *Comparative Political Studies* 40, no.2 (2007): 122–144.

Mann, James. *The China Fantasy: How Our Leaders Explain Away Chinese Repression*. New York: Viking, 2007.
Mann, Michael. *The Sources of Social Power: A History of Power from the Beginning to A.D. 1790* 1. Cambridge: Cambridge University Press, 1986.
―― "The Autonomous Power of the State: Its Origins, Mechanisms, and Results." *European Archive of Sociology* 25 (1984): 185–213.
Mao Zedong. *Mao Zedong and the Political Economy of the Border Region: A Translation of Mao's Economic and Financial Problems*. Translated by Andrew Watson. Cambridge: Cambridge University Press, 1980.
Marson, James. "Russia and China in Major Natural-Gas Supply Pact." *The Wall Street Journal*, March 22, 2013.
Mattlin, Mikael. "The Chinese Government's New Approach to Ownership and Financial Control of Strategic State-Owned Enterprises." *BOFIT Discussion Papers* 10 (2007).
Meidan, Michal, Phillip Andrews-Speed and Ma Xin. "Shaping China's Energy Policy: Actors and Processes." *Journal of Contemporary China* 18, no.61 (2009): 591–616.
Migdal, Joel. *Strong Societies and Weak States: State-Society Relations and State Capabilities in the Third World*. Princeton, New Jersey: Princeton University Press, 1988.
Miller, Alice L. "Institutionalisation and the Changing Dynamics of Chinese Leadership Politics." In *China's Changing Political Landscape*, edited by Cheng Li. Washington D.C.: Brookings Institution Press, 2008a: 61–79.
―― "The CCP Central Committee's Leading Small Groups." *China Leadership Monitor* 26 (2008b).
Miller, Leland. "In Search of China's Energy Authority." *Far Eastern Economic Review* 169, no.1 (2006): 38–42.
Mirsky, Jonathan. "China: Politics as Warfare." *The New York Review of Books*, June 21, 2012.
MOFCOM. "China-Africa Economic and Trade Cooperation (1)." Ministry of Commerce, People's Republic of China, February 15, 2011.
Morgan, Maria Chan. "Administrative Reforms in China." In *Comparative Bureaucratic Systems*, edited by Krishna K. Tummala. Lexington: Lexington Books, 2003.
Mouawad, Jad. "Warning on Impact of China and India Oil Demand." *The New York Times*, November 7, 2007.
Nathan, Andrew J. "China's Political Trajectory: What are the Chinese Saying?" In *China's Changing Political Landscape*, edited by Cheng Li. Washington D.C.: Brookings Institution Press, 2008: 25–43.
―― "China's Resilient Authoritarianism." *Journal of Democracy* 14, no.1 (2003): 6–17.
Naughton, Barry. "China's Distinctive System: Can It be a Model for Others?" *Journal of Contemporary China* 19, no.65 (2010): 437–460.
―― "A Political Economy of China's Economic Transition." In *China's Great Economic Transformation*, edited by Loren Brandt and Thomas G. Rawski. Cambridge: Cambridge University Press, 2008a: 91–135.
―― "SOE Policy: Profiting the SASAC Way." *China Economic Quarterly* 12, no. 2 (2008b): 19–26.

—— "China's Left Tilt: Pendulum Swing or Midcourse Correction?" In *China's Changing Political Landscape*, edited by Cheng Li. Washington D.C.: Brookings Institution Press, 2008c: 142–158.

—— *The Chinese Economy: Transitions and Growth*. Cambridge, Massachusetts: The MIT Press, 2007.

—— "Top-Down Control: The SASAC and the Persistence of State Ownership in China." Paper presented at the conference on "China and the World Economy" Leverhulme Centre for Research on Globalisation and Economic Policy (GEP), University of Nottingham, June 23, 2006.

—— *Growing Out of the Plan: Chinese Economic Reform, 1978–1993*. New York: Cambridge University Press, 1996.

—— "Hierarchy and the Bargaining Economy: Government and Enterprise in the Reform Process." In *Bureaucracy, Politics, and Decision Making in Post-Mao China*, edited by Kenneth Lieberthal and David Lampton. Berkeley: University of California Press, 1992.

Naughton, Barry and Dali L. Yang. "Holding China Together: An Introduction." In *Holding China Together: Diversity and National Integration in the Post-Deng Era*, edited by Barry Naughton and Dali L. Yang. Cambridge: Cambridge University Press, 2004.

NDRC. *China's Medium and Long Term Energy Conservation Plan*. Beijing, November 2004.

Ng, Eric. "Beijing Weighs Up Tough Choices in Balancing Act on Energy." *South China Morning Post*, December 27, 2007.

—— "Oil Refiners Bow to Pressure on Output Rise." *South China Morning Post*, November 20, 2007.

—— "Balancing Refiners' Concerns Against Public's Interest a Tough Act." *South China Morning Post*, August 23, 2005.

North, Douglass. *Structure and Change in Economic History*. New York: Norton, 1981.

O'Donnell, Guillermo. *Modernisation and Bureaucratic-Authoritarianism: Studies in South American Politics*. Berkeley: Institute of International Studies, University of California, 1973.

OECD. *OECD Economic Surveys: China 2010*. Paris: OECD, 2010.

—— *China: Defining the Boundary between the Market and the State*. OECD Reviews of Regulatory Reform. Paris: OECD, 2009.

—— *China 2008: Encouraging Responsible Business Conduct*. Paris: OECD, 2008a.

—— *OECD Investment Policy Reviews: China 2008*. Paris: OECD, 2008b.

—— *OECD Code for Governance of State Owned Enterprises*. Paris: OECD, 2005.

—— *China's Worldwide Quest for Energy Security*. Paris: OECD/IEA, 2000.

Oi, Jean. "The Role of the Local State in China's Transitional Economy." *The China Quarterly* 144 (1995): 1132–1149.

Oil & Gas Journal. "China's Oil Price Reforms a Major Step in Deregulating its Petroleum Sector" 96, no.32 (1998).

ONELG/NDRC. "Office of the National Energy Leading Group." National Development and Reform Commission (NDRC), People's Republic of China, 2011. http://en.ndrc.gov.cn/mfod/t20060829_82145.htm

Oster, Shai and Spender Schwartz. "Beijing Disputes IEA Data on Energy." *The Wall Street Journal*, July 21, 2010.

Page, Jeremy, Wayne Ma and Brian Spegele. "China Probes Former Oil Company Head." *The Wall Street Journal*, September 1, 2013.
Pearson, Margaret. "Governing the Chinese Economy: Regulatory Reform in the Service of the State." *Public Administration Review* 67, no.4 (2007): 718–730.
—— "The Business of Governing Business in China: Institutions and Norms of the Emerging Regulatory State." *World Politics* 57 (2005): 296–322.
Pei, Minxin. "China's Iran Dilemma." *BBC*, January 20, 2012.
—— "China's Big Energy Dilemma." *The Straits Times*, April 13, 2006a.
—— *China's Trapped Transition: The Limits of Developmental Autocracy.* Cambridge, Massachusetts: Harvard University Press, 2006b.
—— "Contradictory Trends and Confusing Signals." *Journal of Democracy* 14, no.1 (2003): 73–81.
Perry, Elizabeth and Christine Wong. *The Political Economy of Reform in Post-Mao China.* Cambridge and London: Harvard University Asia Center, 1985.
People's Daily. "China Develops 5 Trillion Yuan Alternative Energy Plan." July 22, 2010a.
—— "China's Oil Production Ranks Fourth in the World." February 9, 2010b.
PetroChina. *PetroChina Securities and Exchange Commission Form 20-F 2005–2006.* Washington D.C.: United States Securities and Exchange Commission, 2007.
"PetroChina Share Price a Victim of Speculation, Government Control." *ChinaStakes*, September 1, 2008: http://www.chinastakes.com/2008/11/petrochina-share-price-a-victim-of-speculation-government-control.html
"PetroChina, Sinopec to Receive Subsidies." *China Economic Review*, April 21, 2008.
Polidano, Charles. "Don't Discard State Autonomy: Revisiting the East Asian Experience of Development." *Political Studies* 49, no.3 (2001): 513–527.
Ragin, Charles. *Fuzzy-Set Social Science.* Chicago: Chicago University Press, 2000.
Ramo, Joshua. *The Beijing Consensus.* London: Foreign Policy Center, 2004.
Ren, Daniel. "Reshuffle Puts CNOOC Chairman Into Top Slot at Oil Refinery Sinopec." *South China Morning Post*, April 9, 2011.
Reuters. "China Crude Stocks Near IEA Member Level." *Reuters*, July 3, 2009.
"Reviewing China's Energy Policies." *Beijing Review*, January 14, 2008.
Rice, Xan. "China's Economic Invasion of Africa." *The Guardian*, February 6, 2011.
Rodrik, Dani. *One Economics, Many Recipes: Globalization, Institutions, and Economic Growth.* Princeton: Princeton University Press, 2007.
Romero, Simon and Alexei Barrionuevo. "Deals Help China Expand Sway in Latin America." *The New York Times*, April 15, 2009.
Rosen, Daniel and Trevor Houser. *China Energy: A Guide for the Perplexed.* Washington D.C.: Peterson Institute for International Economics, 2007.
Ross, Robert. "Beijing as a Conservative Power." *Foreign Affairs* 76, no.2 (1997): 33–44.
Rotberg, Robert I. *China into Africa: Trade, Aid, and Influence.* Washington D.C.: Brookings Institution Press, 2008.
SASAC. State-Owned Assets and Administration Commission of the State Council: *Central SOEs*, http://www.sasac.gov.cn/n2963340/n2971121/n4956567/4956583.html
—— and Yan Hairong. "Friends and Interests: China's Distinctive Links with Africa." *African Studies Review* 50, no.3 (2007): 75–114.

Sauvant, Karl P. "New Sources of FDI: The BRICs." *The Journal of World Investment and Trade* 6, no.5 (2005): 639–709.
Shambaugh, David. *China's Communist Party: Atrophy and Adaptation.* Washington D.C.: Woodrow Wilson Center Press, 2008.
Shieh, Shawn. "Corruption, Economic Growth and Regime Stability in China's Peaceful Rise." In *China's 'Peaceful Rise' in the 21st Century: Domestic and International Conditions*, edited by Sujian Guo. Aldershot, UK: Ashgate, 2006.
Sheives, Kevin. "China Turns West: Beijing's Contemporary Strategy Towards Central Asia." *Pacific Affairs* 79, no.2 (2006): 205–224.
Shichor, Yitzhak. "Sudan: Neo-Colonialism with Chinese Characteristics." In *China in Africa*, edited by Arthur Waldron. Washington D.C.: The Jamestown Foundation, 2008.
Shirk, Susan. *China: Fragile Superpower.* New York: Oxford University Press, 2007.
—— "The Chinese Political System and the Political Strategy of Economic Reform." In *Bureaucracy, Politics, and Decision Making in Post-Mao China*, edited by David Lampton and Kenneth Lieberthal. Berkeley: University of California Press, 1992.
Skocpol, Theda. "Bringing the State Back In." In *Bringing the State Back In*, edited by Peter B. Evans, Dietrich Rueschemeyer and Theda Skocpol. Cambridge: Cambridge University Press, 1985.
Smyth, Helen. "China's Petroleum Industry." *Far Eastern Survey* 15, no.2 (1946): 187–190.
Soifer, Hillel and Matthias vom Hau. "Unpacking the Strength of the State: The Utility of State Infrastructural Power." *Studies in Comparative International Development* 43, no. 3–4 (2008).
Spence, Jonathan. *The Search for Modern China*, 2nd ed. New York: W. W. Norton & Company, 1999.
—— *To Change China: Western Advisers in China, 1620–1960.* New York: Penguin Books, 1980.
Stanaway, David and Benjamin K. Lim. "China Mulls New Energy 'Super-Ministry'." *Reuters*, January 6, 2012.
Stanislaw, Joseph and Daniel Yergin. "Oil: Reopening the Door." *Foreign Affairs* 72, no.4 (1993): 81–93.
State Council Information Office of China. China Energy White Paper, "China's Energy Conditions and Policies." Beijing (2007).
Stratfor. "At the Peak of China's Oil Dilemma." January 20, 2012.
—— "China: An Energy Superministry?" January 28, 2010.
—— "China: The New Energy Bureau Emerges." July 30, 2008a.
—— "China: The Energy Bureau Moves Against its Parent." June 2, 2008b.
—— "Duma: Keep Slavneft in Russian Hands." December 16, 2002.
Taylor, Ian. "China's Rising Presence in Africa." *China Review International* 16, no.2 (2009): 155–159.
—— "China's Oil Diplomacy in Africa." *International Affairs* 82, no.5 (2006): 937–959.
Tenev, Stoyan and Chunlin Zhang. *Corporate Governance and Enterprise Reform in China: Building the Institutions of Modern Governance.* Washington D.C.: The World Bank and the International Finance Corporation, 2002.
The Economist. "Let a Million Flowers Bloom." 398, no.8724 (2011): 79.
—— "China Buys Up the World." 397, no.8708 (2010a): 11.

—— "Picking Winners, Saving Losers." 396, no.8694 (2010b): 68–70.
—— "The New Colonialists." 386, no.8571 (2008a): 13.
—— "No Strings." 386, no.8571 (2008b): 14–17.
—— "A Ravenous Dragon." 386, no.8571 (2008c): 3.
—— "China's Gas Guzzler." 375, no.8432 (2005a): 85.
—— "From T-shirts to T-bonds." 376, no.8437 (2005b): 65–67.
—— "Chinese Industry and the State: The Myth of China Inc." 376, no.8442 (2005c).
—— "Getting Their Skates On: Long Overdue Reform of China's Capital Markets is in the Air." 355, no.8173 (2000): 72.
"The 11th Five Year Plan: Targets, Paths and Policy Orientation." *Gov.cn: The Chinese Government's Official Web Portal*, March 23, 2006.
The New York Times. "Op-Ed: Google and China." March 23, 2010: A26.
Tilly, Charles. *Democracy*. Cambridge: Cambridge University Press, 2007.
—— "Warmaking and State Making as Organized Crime." In *Bringing the State Back In*, edited by Peter B. Evans, Dietrich Reuschemeyer and Theda Skocpol. New York: Cambridge University Press, 1985.
Tomasic, Roman and Neil Andrews. "Minority Shareholder Protection in China's Top 100 Listed Companies." *Australian Journal of Asian Law* 9, no.1 (2007): 88–119.
Troner, Alan. "China's Oil Sector: Trends and Uncertainties." The Rise of China and its Energy Implications, Energy Forum of the James A. Baker III Institute for Public Policy of Rice University, 2011.
Tsai, Kellee S. *Capitalism Without Democracy: The Private Sector in Contemporary China*. Ithaca, New York: Cornell University Press, 2007.
Tu, Jianjun. "China's New Energy Commission and Energy Policy." *China Brief* 8, no.7 (2008).
US-China Economic and Security Review Commission, *2005 Report to Congress*. Washington D.C.: US Government Printing Office, 2005.
US-China Security Review Commission, Hearing on China's Future Energy Development and Acquisition Strategies, "Statement by Dr Gal Luft." July 21, 2005.
United States Trade Representative Office. *2010 Report to Congress on China's WTO Compliance*. Washington D.C.: US Government Printing Office, 2010.
Vivoda, Vlado. "China Challenges Global Capitalism." *Australian Journal of International Affairs* 63, no.1 (2009): 22–40.
Vivoda, Vlado and James Manicom. "Oil Import Diversification in Northeast Asia: A Comparison between China and Japan." *Journal of East Asian Studies* 11, no.2 (2011): 223–254.
Wade, Robert. *Governing the Market: Economic Theory and the Role of Government in East Asian Industrialization*. Princeton, New Jersey: Princeton University Press, 1990.
Walder, Arthur. "The Party Elite and China's Trajectory of Change." *China: An International Journal* 2, no.2 (2004).
Walter, Carl E. and Fraser J. T. Howie. *Red Capitalism: The Fragile Financial Foundation of China's Extraordinary Rise*. Singapore: Wiley, 2011.
Wang, Haijiang H. *China's Oil Industry and Market*. Amsterdam, Oxford: Elsevier, 1999.
—— "China's Oil Policy and Its Impacts." *Energy Policy* 23, no.7 (1995): 627–635.

Wang, Shaoguang. "The Rise of the Regions: Fiscal Reform and the Decline of Central State Capacity in China." In Andrew Walder (ed.) *The Waning of the Communist State: Economic Origins of Political Decline in China and Hungary*. Berkeley: University of California Press, 1995: 87–114.
—— "Building a Strong Democratic State: On Regime Type and State Capacity." Papers of the Center for Modern China, Number 4, February 1991.
Wang Shaoguang and Hu Angang. *The Chinese Economy in Crisis: State Capacity and Tax Reform*. New York: East Gate Book, 2001.
Weisman, Jonathan. "China's Demands Anger Congress, May Hurt Bid." *The Washington Post*, July 6, 2005, D01.
Weiss, Linda. *The Myth of the Powerless State*. Ithaca, New York: Cornell University Press, 1998.
Wildau, Gabriel. "Enterprise Reform: Albatross Turns Phoenix." *China Economic Quarterly* 12, no.2 (2008): 27–41.
Woetzel, Jonathan R. "Reassessing China's State-Owned Enterprises." *Forbes.com*, August 7, 2008.
Wolf, Martin. *FT Panel: Davos Digest*, audio file (2008): http://www.ft.com/intl/cms/31ef7012-d33c-11dc-b861-0000779fd2ac.mp3
Woodard, Kim. *The International Energy Relations of China*. California: Stanford University Press, 1980.
Wu, Jinglian. *China's Long March Toward a Market Economy*. San Francisco, California: Long River Press, 2005.
Xin, Ma and Phillip Andrews-Speed. "The Overseas Activities of China's National Oil Companies: Rationale and Outlook." *Minerals and Energy* 21, no.1 (2006): 17–30.
Xinhua. "Zhang Yi Becomes Head of SASAC." *China.org.cn*, December 25, 2013.
—— "National Energy Commission was Established." January 27, 2010a: http://news.xinhuanet.com/fortune/2010-01/27/content_12886602.htm
—— "China-Africa Trade to Top 100 Billion USD Again This Year." *Xinhua*, October 3, 2010b.
—— "Chinese Maritime Authority Prepared to Sue ConocoPhillips Over Oil Spills." *Xinhua*, August 25, 2008.
Xu, Ting. "Destination Unknown: Investment in China's 'Go Out' Policy." *China Brief* 11, no.17 (2011).
Yam, Shirley. "Corporate Governance Takes Back Seat in Musical Chairs." *South China Morning Post*, April 16, 2011.
Yan, Zhou. "Energy Fuels Booming Bilateral Relationship." *China Daily*, June 17, 2011.
Yang, Dali. *Remaking the Chinese Leviathan: Market Transition and the Politics of Governance in China*. Stanford, California: Stanford University Press, 2004.
—— "State Capacity on the Rebound." *Journal of Democracy* 14, no.1 (2003): 43–50.
Yang, Dali and Yanzhong Huang. "Bureaucratic Capacity and State-Society Relations in China." *Journal of Chinese Political Science* 7, no.1–2 (2002): 19–46.
Yeo, Yukyung. "Remaking the Chinese State and the Nature of Economic Governance? The Early Appraisal of the 2008 'Super-Ministry' Reform." *Journal of Contemporary China* 18, no.62 (2009a): 729–743.
—— "Between Owner and Regulator: Governing the Business of China's Telecommunications Service Industry." *The China Quarterly* 200 (2009b): 1013–1032.

Yergin, Daniel. *The Quest: Energy, Security and the Remaking of the Modern World*. London: Penguin books, 2011.
—— "Ensuring Energy Security." *Foreign Affairs* 85, no.2 (2006): 69–82.
—— *The Prize: The Epic Quest for Oil, Money and Power*. New York: Simon and Schuster, 1991.
—— "Energy Security in the 1990s." *Foreign Affairs* 67, no.1 (1988): 110–132.
Yergin, Daniel, Dennis Eklof and Jefferson Edwards. "Fueling Asia's Recovery." *Foreign Affairs* 77, no.2 (1998): 34–50.
Yongnian, Zheng. *Will China Become Democratic? Elite, Class and Regime Transition*. Singapore: Eastern University Press, 2004.
Yoshihara, Toshi and James R. Holmes. "China's Energy-Driven 'Soft Power'." *Orbis* 52, no.1 (2008): 123–138.
Yu, Wang. "A Treasure Trove Called Bohai Bay." *China Daily*, July 9, 2007.
Zha, Daojiong. "China's Energy Security." *Survival* 48, no.1 (2006): 179–190.
Zhang, Allan. "Going Abroad – China's Corporations Go Global." *PricewaterhouseCoopers Report* (2004): 1–11.
Zhang, Hong and Eric Ng. "Corruption Probe Raise State-Owned Enterprise Reform Hopes." *South China Morning Post*, September 3, 2013.
Zhang, Kenny. "Going Global: The Why, When, Where and How of Chinese Companies' Outward Investment Intentions." *Canada-Asia Agenda* 5 (2005).
Zhang Libin and Jason Lee. "Situation Report: Energy Policy." *China Security: A Journal of China's Strategic Development* 11 (2008).
Zhang, Shu Guang. *Economic Cold War: America's Embargo against China and the Sino-Soviet Alliance 1949–1963*. Stanford, California: Stanford University Press, 2001.
Zhang, Yongjin. *China's Emerging Global Businesses: Political Economy and Institutional Investigations*. New York: Palgrave Macmillan, 2003.
Zhao, Dingxin. "The Mandate of Heaven and Performance Legitimacy in Historical and Contemporary China." *American Behavioral Scientist* 53 (2009): 416–433.
Zhao, Suisheng. "China's Global Search for Energy Security: Cooperation and Competition in the Asia-Pacific." *Journal of Contemporary China* 17, no.55 (2008): 207–227.
—— "The Structure of Authority and Decision-Making: A Theoretical Framework." In *Decision-Making in Deng's China: Perspectives from Insiders*, edited by Carol Lee Hamrin and Suisheng Zhao. Armonk, New York: M. E. Sharpe, 1995.
Zheng, Yongnian. *De Facto Federalism in China: Reforms and Dynamics of Central-Local Relations*. Singapore: World Scientific Publishing, 2007.
—— *Globalization and State Transformation in China*. Cambridge: Cambridge University Press, 2004.
Zhu, Yuchao. "'Performance Legitimacy' and China's Political Adaptation Strategy." *Journal of Chinese Political Science* 16, no.2 (2011): 123–140.
Ziegler, Charles E. "The Energy Factor in China's Foreign Policy." *Journal of Chinese Political Science* 11, no.1 (2006): 1–23.
Zweig, David and Bi Jianhai. "China's Global Hunt for Energy." *Foreign Affairs* 84, no.5 (2005): 25–38.

Index

adaptive capacity, 34, 35
adaptive governance, 35, 44
Addis Ababa Action Plan, 171
administrative reform, 105, 114, 118, 121, 127
Africa, Chinese trade with, 147, 166, 168, 170, 171, 184
'African policy', 171
Angarsk pipeline, 130
Asian Financial Crisis in 1997, 47, 119
authoritarian political systems, 179
'authoritarian resilience', 180
authoritarian state capacity in liberal world order, 175
   implications of China's state-led oil strategies for business and politics, 182–7
   rise of China's market authoritarianism model, 178–82
authority, political and bureaucratic hierarchies of, 14–17

'big battle' formula, 79
'Big Push industrialisation', 70
Bijian, Zheng, 6
blue water navy, development of, 11
Breakneck economic growth, 6
BRICS, 186
bureaucratic authoritarianism (BA), 3, 16, 66–9
bureaucratic organisation, 2, 13
bureaucratic reform, 105–7
bureaucratic structure of authority, 15

capacity building, 37, 49, 56, 132
capacity-building efforts, 22, 108
career incentive structures, 3
CDB (Central Development Bank), 168, 169
censorship, 183

Central Asian energy cooperation, 170
centralised energy agencies, 140–1
central leadership, 45, 67
Central Organisation Department (COD) of CCP, 64, 161
Central Party School, 63
central party-state, 151
Central Work Conference on Foreign Affairs, 167
Chaebol, 14
Charles de Gaulle, 81
Chen Yuan, 169
China-Africa cooperation, 171–2
China Eximbank, 168, 169
China Mobile, 126
China National Chemicals Import and Export Corporation (Sinochem), 94, 95
China National Energy Strategy and Policy 2020, 132
China Petrochemical Corporation, 94
China Telecom, 126
'China threat theory', 31
China Unicom, 126
China's Trapped Transition, 42
Chinese Communist Party (CCP), 3, 16, 21, 27, 29, 45, 56–7, 76, 155, 177, 178, 180
   authority and legitimacy, strengthening, 61–6
   bureaucratic authoritarianism, 66–9
   Central Organisation Department (COD) of, 64
   and economic reform, 126
   neglected role of, 59–61
   'performance dilemma', 62
Chinese People's Political Consultative Conference, 167
CNOOC (China National Offshore Oil Corporation), 91, 93, 95, 96, 98, 105, 129, 148, 157–8, 163–4

207

CNPC (China National Petroleum Corporation), 12, 14, 94, 95, 96, 97, 98, 103, 105, 113, 128, 129, 157–8, 166, 168
coercive capacity, 35
'co-governance structure', 110
collaboration governed by hierarchy, 178
command economic model, 70
Commission of Science Technology and Industry for National Defense (COSTIND), 137
comprehensive tax, 45
conceptualisation of energy security, 17–20
'conservation-minded society', 132
'contested modernity', 186
Coordinating Committee for Multilateral Export Control (COCOM), 76
corporate governance, 151
  with Chinese characteristics, 153–6
corporatisation, 28, 93, 129
corrupt practices, 159
crude oil
  imports, 11
  price reform, 100
  pricing, 103
Cultural Revolution, 76, 82–4

Daqing method, 71, 74, 78, 79, 80, 82, 92
decentralisation, 41, 51, 58, 88, 101, 105–7, 110, 120, 129
decision-making, 15
demand for oil, 5–6
demand-side management, 18, 19
Deng Xiaoping, 88, 101
Deng's axiom, 51
department for energy cooperation, 138
despotic power, 38
developing countries, China's growing investments in, 186
diplomatic efforts, 169–70
'disintegration' thesis, 41
diversification strategy, 10–11
domestic and international oil policy behaviour, 176

domestic crudes, 103
domestic oil production, 73–4
  cost of, 10
downstream oil prices, 97
dual-track system, 99, 108
Dushanzi oil field, 73

economic crisis of 1960 to 1961, 75
economic decision-making, 56, 57
economic planning, influence of oil industry on, 83
economic reform, 6
  since the mid-1990s, 121
'economic rights', 170
economic stimulus package, 7
economy of China in mid-1970s, 72
elite policy agendas, 55
elite politics, incremental institutionalisation of, 44
'embedded autonomy', 39
'energy-backed loans' (EBLs), 169
Energy Bureau, 133–4, 137
energy challenges, 114
energy development, 71
energy governance in China, 176
energy independence, 78
  under conditions of economic autarky, 71
energy intensity
  and carbon dioxide emissions, reductions in, 146
  reducing, 19
energy policymaking
  FA perspective on, 53–9
energy security
  Beijing's evolving conceptualisation of, 17–20
  of China, 177
  conceptualisation of, 17–20
'external posture', 32

financial crisis, in US, 139
first and second Reform Eras, distinction between, 52
first Reform Era, 22
  China's oil industry in, 107–11
fiscal capacity, 35, 104, 181
fiscal extractive capacity, 35
fiscal reforms, 45

Index 209

fledgling regulatory state, 114, 115
Foreign Direct Investment (FDI), 88
foreign equity investments, 18
foreign oil dependency, 8
    pitfalls of, 73–6
foreign oil technologies, purchases of, 85
foreign participation in oil sector development, 90–2
Forum on China-Africa Cooperation (FOCAC), 171–2
fossil fuel consumption, 8, 9
fragmentation, 28, 50, 54, 58
'fragmentation thesis', 31, 41
fragmented authoritarianism (FA) model, 2, 14–15, 27, 51, 55, 176, 186
France, 19, 20, 84
'free wheeling' 1980s, 118
Fu Chengyu, 162, 163

global financial crisis (GFC) in 2008, 47
global steel production, 7
'go global' policy, 128, 133, 152, 166–9
'go international', 128
Google, 183
government-NOC relationship, 178
government restructuring since 1998, 118–27
'Go West' campaign, 161
gradualism, 22, 23, 142
    advantages of, 108
    flipside of, 108
gradualist, 177
    reform, 42–3
'grasp the large, release the small' policy, 13, 114, 119, 123, 133
Great Leap Forward, 70, 71, 76–7
'growing out of the plan' strategy, 120
'guerrilla-style policymaking', 44

'harmonious society', 124
head-to-head competition among firms, 165
"horizontal interorganisational bargaining", 2

Hu Jintao, 18, 38–9, 132, 170, 172
Hu-Wen leadership, 18, 19, 62, 124, 144
Hu Yaobang, 60

IMF, 183, 184
imperative of taxation, 35
incremental institutional change, 142
"industry-wide subsidy", 104
infrastructural power, 38
innovative capacity, 35
institutional arrangements in China, 176, 177
institution building, 43–4
integration of China within US-led international system, 183
International Energy Agency (IEA), 5
international oil companies (IOCs), 5, 149–50
interplay of elite, 50
IPOs (Initial Public Offering), 156
    financing, 97
Iran, 184–5
Iraq War, 112, 130, 131, 143

Japan, 76, 84
Jiang Jiemin, 14, 118, 123, 143, 102
job swaps, 165

Kazakhstan, 171
Khrushchev, 75
Korean economy, 14

leadership in China, 173
'Learn from Daqing', 71
learning capacity, 34, 108
legitimacy, performance-based, 62
legitimation capacity, 35
Li Keqiang, 139
liberal market approaches, 9
'limited and managed' competition, 126
Lin Boqiang, 139
'loans-for-oil' deals, 168–9

"managed competition", establishing, 125
managerial incentives, 65
Mao era, 175

Maoist innovations, 70
Maoist system, 22, 86
Mao Zedong, 70, 74
Mao Zedong Thought, 72, 79
Mao's dictatorial rule, 2
market authoritarianism model of China, 178–82
market-oriented economy, 22, 119
market-refined products, 100
market transition
　oil industry and firm development under, 22–5
Marxist theory, 62, 63
mercantilist approach of China, 182
micro-control, 122, 125
Microsoft, 183
Ministry of Energy (MOE), 105–6
Ministry of Foreign Economic Relations and Trade (MOFERT), 91
Ministry of Foreign Trade and Economic Cooperation (MOFTEC), 91
Ministry of Geology (MOG), 83
Ministry of Petroleum Industry (MPI), 74, 82, 83, 88, 128
'modern enterprise system', 119
multilateral cooperation, 170

'national champions', 114, 119, 149, 153
National Development and Reform Commission (NDRC), 16, 90, 115, 124, 126, 133, 138, 139, 143
'national economy supreme command', 82
National Energy Administration (NEA), 137, 138, 139
National Energy Commission (NEC), 107, 138–41
National Energy Leading Group (NELG), 134, 137
national oil companies (NOCs), 1, 2, 3, 5, 21, 24, 27, 29, 88, 89, 110, 111, 117, 139, 148, 149
　and central government, 150
　corporate governance with Chinese characteristics, 153–6
　creation of, 92–8
　'go global' policy, 166–9
　leadership, 161–6, 173
　oil diplomacy, 169–74
　ownership and regulation of, 156–61
national self-reliance, 70, 71
'neomercantilist' approach, 1, 166
New York Stock Exchange (NYSE), 158
NOCs, 58
　and Chinese government, 178, 185
nomenklatura system, 65, 161–2
'no political strings attached' investment strategy, 184
North Atlantic Treaty Organisation (NATO) countries, 76

Office of the NELG (ONELG), 135
offshore contracting, 91
oil demand between 2009 and 2035, 5–6
oil demand in 2004, 112
oil diplomacy, 169–74
oil exports, 72
oil governance regime, 121
　recentralising, 127–43
oil import, 147
　dependency, 8–9
oil industry reform (2003–2013), 117
oil policy, 3
　approach, 8–11
　formulation and implementation, 113, 143–8
　socioeconomic dimensions of, 20–2
oil policymaking, 116
oil prices, 8, 20, 88–9
oil-rich countries, 171
oil sector development, 88, 89
　bureaucratic reform and decentralisation, 105–7
　creation of China's NOCs, 92–8
　foreign participation in, 90–2
　in first Reform Era, 107–11
　oil price reform conundrum, 98–105
oil security, 131
oil security dilemma, 4–8
oil self-sufficiency
　under Mao, 77–81

Index 211

'oil shock treatment', 101–2
oil state capacity, building, 12–14
oil state capacity, rebuilding, 112
  government restructuring since
    1998, 118–27
  oil policy formulation and
    implementation, 143–8
  recentralising oil governance
    regime, 127–43
oligopolistic competition, 125
OPEC oil embargo, 85
outward foreign direct investment
  (OFDI), 167, 168
overseas investments, 128, 168

party-state control, 89–90
Patrick Cheung, 101
"peaceful foreign policy", 94
People's Bank of China (PBC), 68
People's Republic of China (PRC), 20,
  25, 70
performance-based goals, 62
PetroChina, 98, 158, 159, 161
'petro purge', 24, 160, 173
polar characterisations of China, 48
policy compliance, 96
policy inertia, 50
policymaking, in China, 32
policymaking, party-state centred
    explanation of, 1
  authority, political and
    bureaucratic hierarchies of,
    14–17
  Beijing's evolving conceptualisation
    of energy security, 17–20
  China's oil security dilemma, 4–8
  oil industry and firm development
    under market transition, 22–5
  oil policy, socioeconomic
    dimensions of, 20–2
  oil policy approach, 8–11
  oil state capacity, building, 12–14
  research methods, 25–6
policy sectors, in China, 17
political future of China, 179
political loyalty, 65
political system of China, 31, 52
post 9/11 terrorist attacks, 130
post-GFC China, 24

post-Mao industrial development,
  175
post-Tiananmen period, 16
power redistribution, 67
Primary Energy Consumption in
  2012, 7
production-sharing agreements
  (PSAs), 90–1

Qing government, 73
quantitative methods, 26
quasi-military mass campaign model,
  85

Red Guard activists, 82
refined oil products, 100
'reform and opening', 88
Reform Era marketisation, 87
Reform Era oil industry reform, 2,
  175, 187
reform strategies, weaknesses of, 42
'reform without losers', 108
regulatory state, building, 118–27
renewable energy development, 146
research methods, 25–6
'resilient authoritarianism' thesis, 44
'rise of China', 31
'Russian chauvinism', 75
Russia's Arctic shelf, 19

Scientific Development Concept, 18,
  63–4, 132, 144
SDPC (State Development Planning
  Commission), 90, 103
'second era of reform', 15
second Reform Era, 59, 176–7
'second revolution', 88
sectoral governance, 31
securitisation of energy issues, 112
self-reliance, 18
  and economic autarky, 72
  pursuit of, during Great Leap
    Forward, 76–7
SETC (State Economic and Trade
  Commission), 106
Shambaugh's thesis, 45, 46
Shanghai Cooperation Organisation
  (SCO), 170
share ownership, 157

Shengli, 92
'single-shot' hypothesis testing, 26
Sino-American confrontation, 81
Sinopec, 93–8, 129, 148, 154, 157, 158, 159, 163, 173
Sino-Soviet alliance, 74–5
Sino-Soviet rift, 71, 75–6
Sixth Five-Year Plan (1981–5), 90
slow shift away from economic autarky, 84–7
small groups, leading, 65
'small planning commission', 82
socialist era of oil self-sufficiency, 70
　foreign oil dependency, 73–6
　pursuit of self-reliance during Great Leap Forward, 76–7
　realisation of oil self-sufficiency under Mao, 77–81
　slow shift away from economic autarky, 84–7
　third front and Cultural Revolution, 81–4
socialist market economy, 50
social responsiveness, 44
Soviet Union, 75
　significant dependency on, 71
'the state,' development of, 59–60
State Asset Supervision and Administration Commission (SASAC), 13, 14, 28, 123, 125, 148, 149, 155, 156–7, 162, 168, 173
state capacity, 12–14, 31, 89
　and autonomy, 37–9
　concept of, 34–7
　contrasting views of, 39–49
　traditional statist conceptualisations of, 39
　utility as an analytical tool, 36
state capitalist model of China, 181, 183
State Development Planning Commission (SDPC), 122, 123, 133
State Economic and Trade Commission (SETC), 14, 121–2, 123, 133, 143
State Energy Office (SEO), 135

state-led oil strategies of China, 150
　for business and politics, 182–7
state-led policy approaches, 10
state-owned enterprises (SOEs), 13, 14, 24, 53, 65, 93, 95–6, 114, 116, 119, 120–3, 125, 149, 151, 153, 155, 156, 157, 158, 160, 161–2, 165, 168, 181
State Petroleum and Chemical Industry Bureau (SPCIB), 106, 107
steering/regulatory capacity, 35
strategic petroleum reserve (SPR), 11
strategic sectors, 149
Sudan, China's oil dealings with, 184
super-ministries, creation of, 124–5
'super-ministry' reform, 124
Su Shulin, 159

taxation powers, 35
Tehran, 184
third front development, 81–3
Tiananmen massacre, 61
Tiananmen Square, 39
trait of Malacca, 6
'transnational operational strategy', 166
Transneft, 130
transportation sector, 8
two-tiered price system, 99

United States
　global steel production, 7
Unocal, 6, 131, 164, 166
upstream and downstream assets, 95
urbanisation, 8

Venezuela, 169
Vietnam War, 81
vis-à-vis NOCs, 185

Wang Jinxi, 79–80, 83
Wang Tao, 128, 129
Wang Yong, 123
weather external shocks, 47
Web-censorship regime, 183
Wen Jiabao, 39, 132, 139
West-East Gas Pipeline Project (WEPP), 152, 160

Wittner, Mike, 10
World Bank, 168, 183, 184
WTO, 183

Xi Jinping, 14, 17, 19, 39, 160
Xinjiang, 74
Xue Feng, 154

Yahoo, 183
Yanchang, 73
Yom Kippur War (1973), 85
Yumen, 73

Zhang Jianyu, 135
Zhang Yi, 123
Zhao Ziyang, 60
Zhongyuan oil field, 92
Zhou Enlai, 72, 80, 83, 85, 90
Zhu Chengzhang, 134
Zhu Rongji, 106, 122, 127, 143, 161, 167
Ziyang, Zhao, 65

GPSR Compliance
The European Union's (EU) General Product Safety Regulation (GPSR) is a set of rules that requires consumer products to be safe and our obligations to ensure this.

If you have any concerns about our products, you can contact us on

ProductSafety@springernature.com

In case Publisher is established outside the EU, the EU authorized representative is:

Springer Nature Customer Service Center GmbH
Europaplatz 3
69115 Heidelberg, Germany

www.ingramcontent.com/pod-product-compliance
Lightning Source LLC
Chambersburg PA
CBHW061806110426
42873CB00042B/47